高等学校土木工程专业“十三五”规划教材
高 校 土 木 工 程 专 业 规 划 教 材

Engineering Geology Practice
工程地质实践

Feng Jinyan, Li Feng, Chen Jun, Jiao Shenhua
冯锦艳　李　峰　陈　军　焦申华　编著

中国建筑工业出版社
CHINA ARCHITECTURE & BUILDING PRESS

图书在版编目（CIP）数据

工程地质实践＝Engineering Geology Practice：英文/冯锦艳等编著．—北京：中国建筑工业出版社，2020.1

高等学校土木工程专业“十三五”规划教材，高校土木工程专业规划教材

ISBN 978-7-112-24444-7

Ⅰ．①工…　Ⅱ．①冯…　Ⅲ．①工程地质-高等学校-教材-英文　Ⅳ．①P642

中国版本图书馆CIP数据核字（2019）第247285号

责任编辑：聂　伟　王　跃
责任校对：芦欣甜

高等学校土木工程专业“十三五”规划教材
高 校 土 木 工 程 专 业 规 划 教 材
Engineering Geology Practice
工程地质实践
Feng Jinyan，Li Feng，Chen Jun，Jiao Shenhua
冯锦艳　李　峰　陈　军　焦申华　编著

*

中国建筑工业出版社出版、发行（北京海淀三里河路9号）
各地新华书店、建筑书店经销
北京红光制版公司制版
北京建筑工业印刷厂印刷

*

开本：787×1092毫米　1/16　印张：8½　字数：207千字
2019年12月第一版　2019年12月第一次印刷
定价：**36.00**元
ISBN 978-7-112-24444-7
（34923）

Book Description

This textbook is a teaching material for the course of engineering geology. The basic practice contents and methods of engineering geology are expounded in this textbook, which is divided into eight chapters, including identification of rock-forming minerals and three kinds of rocks, the application method of the geological compass, field observation methods of geological structures, reading method of geological map, comparisons of engineering geology at home and abroad, appreciation of the world geological wonders and engineering geological practice routes around or in north China.

This textbook is suitable for the undergraduates and post graduates of civil engineering, airport pavement engineering, hydraulic engineering, transportation engineering and other majors, who come from China or other countries, especially from the countries of One Belt and One Road. It is a good reference book for qualifying examinations of professionals, such as civil engineering qualifying examination, geotechnical engineering qualifying examination and environmental assessment qualifying examination in China.

本书为“工程地质学”配套的实习教材，阐述工程地质学实习的基础内容和基本方法。全书共8章，主要包括造岩矿物与三大类岩石的鉴定、地质罗盘仪的使用、地质构造的野外观察与分析、地质图的使用与阅读、中外关于工程地质的主要差异比较、世界著名地质奇迹赏析以及典型路线的地质实习等。

本书可作为高等学校土木工程、机场工程、水利工程、交通运输工程及相关专业的本科生和研究生的教学用书或教学参考书，也可作为留学生，尤其是“一带一路”沿线国家留学生的教学用书。本书也可供参加中国土木工程、岩土工程以及环境评价工程等资格考试的人员参考。

Preface

Engineering geology is one of basic courses for civil engineering, road engineering, airport engineering, hydraulic engineering and other related majors, because it plays an important role in choice of construction site. For decades, with the rapid development of civil engineering at home and abroad, engineering geology has obtained abundant practical experiences and remarkable theoretical results, which provide rich materials for this textbook concerned with those geological factors that influence the location, design, construction and maintenance of engineering works. Accordingly, it draws on a number of geological disciplines such as geomorphology, structural geology, sedimentology, petrology and stratigraphy. In addition, engineering geology involves hydrogeology and some understanding of rock and soil mechanics.

There are eight chapters in this textbook, which can be divided into two parts. The first part, namely from chapter 1 to chapter 6, mainly focus on the basic practical teaching contents, including identification of main rock-forming minerals, identification of three kinds of rocks, the use method of the geological compass, field observation methods of geological structures and reading method of geological map, comparisons of important contents of engineering geology among Chinese standards, Europen standards and American standards. The second part, namely chapter 7 and chapter 8, mainly focus on practical application teaching contents, including appreciation of the world geological wonders and engineering geological practice routes which will be available for teachers and students practice in north China.

This textbook is mainly written for undergraduate and post graduate students of civil engineering. It is hoped that this will also be of value to those involved in the profession.

We regret that this textbook cannot cover all the needs of the variety of readers who may use it. Therefore, references are provided for those who want to pursue some aspects of the subject matter to greater depth. Obviously, students of civil engineering will have done much more reading on engineering geology than the basic geological material. Moreover, any book will reflect the background of its authors and their views of the subject. However, these authors have attempted to give a balanced overview of the subject.

The first, second, third, seventh and eighth chapters are compiled by Feng Jinyan, the fourth chapter is compiled by Li Feng, the fifth chapter is compiled by Chen Jun, and the sixth chapter is compiled by Jiao Shenhua.

The authors gratefully acknowledge all those who have given permission to publish materials from other sources. The authors would like to acknowledge the support of Beihang University.

Feng Jinyan
October 2019

前　言

“工程地质学”是地质工程和土木工程的基础课，也是水利水电工程、交通运输工程等专业的重要必修课，对工程选址具有决定性的作用。几十年来，随着国内外土木工程的飞速发展，工程地质学得到了飞速发展，积累了丰富的实践经验和理论成果，为本教材的编写提供了丰富的素材。因此，本书涉及许多地质学科，如地貌学、构造地质学、沉积学、岩石学和地层学，此外，还包括水文地质学和岩石力学。

本教材共8章，可分为两大部分，第1部分（第1～6章）主要为基础实践教学，包括主要造岩矿物的鉴定、三大类岩石的鉴定、地质罗盘仪的使用、地质构造的野外观察、地质图的阅读，尤其突出了中国规范与欧洲规范以及美国规范关于工程地质重要内容的差异对比。第2部分（第7、8章）主要为实践应用教学，包括世界地质奇迹赏析以及工程地质实习路线，可供在我国华北地区实习的教师和学生使用。

本教材由北京航空航天大学冯锦艳、李峰、陈军以及北京现代金宇岩土工程有限公司焦申华编著，第1～3章以及第7、8章由冯锦艳执笔，第4章由李峰执笔，第5章由陈军执笔，第6章由焦申华执笔。

衷心感谢北京航空航天大学对本教材的支持。

限于作者水平，教材中如有错误和不妥之处，恳请读者批评指正。

冯锦艳

2019年10月

Catalogue

Chapter 1 Identification of common rock-forming minerals

Most of the chemical elements in the Earth's crust exist in the form of compounds, only a few exist in the form of elementary substances. Minerals refer to the natural elements and compounds which possess some chemical compositions and physical properties. The content order of elements in the Earth's crust from high to low is oxygen, silicon, aluminum, iron and other minerals. Rocks can be formed by one or more minerals which are called rock forming minerals.

At present, there are more than 3000 minerals found in nature, most of which are solid. There are over 100 kinds of rock-forming minerals and about 30 kinds of common rock-forming minerals among them, such as orthoclase, anorthose, black mica, white mica, augite, amphibole, talcum, kaolinite, quartz, dolomite, pyrite, magnetite and so on.

1.1 Purposes and requirements

The identification purposes and requirements of the rock-forming minerals are described as follows:

(1) To understand the common physical and chemical properties of rock-forming minerals.

(2) To master the identification methods of rock-forming minerals by naked eyes or simple tools.

(3) To lay a foundation for the identification of common rocks by recognizing the main rock-forming minerals.

1.2 Common tools

The common tools used to identify the rock-forming minerals include magnifying glass, streak plate, knife, magnet, dilute hydrochloric acid, Moh's scale and so on.

1.3 Identification methods of rock-forming minerals

Generally, many methods are required to identify rock-forming minerals comprehensively. All of identification methods should be valid, accurate and fast.

The preliminary appearance observation and identification of minerals is the most convenient and intuitive method. This method is used to identify the physical properties of minerals by naked eyes, tongue and hand, or by knife, magnifying glass, and then to determine the name of the mineral.

In the process of mineral identification, first of all, the color, streak, gloss and transparency should be observed, and then the hardness, cleavage or fracture should be identified. Some minerals have unique properties that can be used to identify minerals, such as the elasticity of mica, the magnetism of magnetite and the smooth feel of talc.

The difference of minerals needs to be identified if their certain property is similar or identical. For example, both calcite ($CaCO_3$) and gypsum ($CaSO_4$) are white in color, but they have different shapes, namely, calcite ($CaCO_3$) is diamond hexahedron, but gypsum ($CaSO_4$) is tabular or fibrous. The most obvious difference is that calcite ($CaCO_3$) produces bubbles when it meets dilute hydrochloric acid, but gypsum ($CaSO_4$) does not.

1.4 Practice contents

1.4.1 Morphological observation of mineral monomers

The morphology of minerals refers to the appearance of minerals, which is the first impression of mineral chemical composition. The same kind of crystalline grains that grow in the same condition always tend to form a particular crystal, which is called crystallization habit. According to the development degree of crystal in three dimensional spaces, it can be divided into three types.

(1) Extended in one direction. The crystal grows in one direction, and the other two directions are poorly developed. It often forms columnar, acicular and threadiness, such as quartz of hexagonal bipyramid or hexagonal prism, hornblende of long columnar or threadiness and andalusite of long columnar.

(2) Extended in two directions. The crystal grows in two directions, and the other direction is poorly developed. It often forms clintheriform, schistose and flaky, such as flaky mica, tabular gypsum and feldspar.

(3) Extended in three directions. The crystal grows in three directions. It often forms cube, rhomb and octahedron, such as garnet, calcite and pyrite of a diamond-shaped rhomb.

1.4.2 Morphological observation of mineral aggregations

In the nature, crystalline minerals rarely appear in monomers, and amorphous minerals do not have regular monomer morphology, so minerals are often identified by the forms of the aggregates.

The aggregate of multiple monomers of the same mineral is called mineral aggregate, and its morphology depends on the morphology and combination styles of monomers. The mineral aggregate is divided into three categories according to the size of the mineral particles or the discernibility as follows:

(1) Phanerocrystalline aggregate whose monomers can be identified by naked eyes.

(2) Cryptocrystalline aggregate whose monomers can be identified by microscope.

(3) Colloidal aggregate whose monomers even cannot be identified by microscope.

1.4.2.1 Phanerocrystalline aggregate morphology

There are four main types of phanerocrystalline aggregate morphology as follows:

(1) Columnar, acicular and fibrous aggregates are mainly composed of mineral particles extended in one direction, such as asbestos, fibrous gypsum, actinolite and andalusite.

(2) Flaky, scaly, tabular aggregates are mainly composed of mineral particles extended in two directions, such as graphite, barite and chlorite.

(3) Granular aggregates are mainly composed of mineral particles extended in three directions, such as olivine and garnet.

(4) Aggregates like crystal cluster refer that mineral granule aggregates that grow in the caves or fissures of the rock wall as basement, such as quartz cluster and calcite crystals.

1.4.2.2 Cryptocrystalline aggregate morphology

Cryptocrystalline aggregate is crystallized directly from a solution or formed by colloform. The aggregates tend to form spheration because of the surface tension of colloids. Cryptocrystalline aggregate often becomes cryptocrystalline or phanerocrystalline to produce a radioactive fibrous structure inside the spheroid after the colloids aging. In addition, cryptocrystalline and colloidal aggregates can also be dense or earthy.

1.4.2.3 Colloidal aggregate morphology

There are 11 types of colloidal aggregates as described below.

(1) Secretinite aggregate is formed due to gradual deposition and filled by the colloid or crystalloid from the cavity wall to the center in a spherical cavity. The characteristic of the aggregate is that most of the constituent substances have the concentric layer structure in the extroversion, and each layer shows strip ring due to different constituents and colors.

When the diameter of secretinite is more than 1cm, it is called the crystalline gland, such as agate. When the diameter of secretinite is less than 1cm, it is called the amygdala, such as the amygdaloid body in volcanic rock.

(2) Concretion aggregate is different from the secretinite aggregate. Its substance grows from the inside out around a center to form an irregular block massive, whose internal structure shows radioactive, concentric layered or compact massive. Some centers are empty and can be filled with other substances. The most common tuberculous form minerals are pyrite, marcasite, limonite, siderite, opal and calcite.

(3) Oolitic and pisolitic aggregates are formed by sedimentation. They have concentric structures that can be formed by growth around a substance, such as sand and mineral fragment. The one that whose size is less than 2mm is oolites aggregate and the one that whose size is more than 2mm is pisolitic aggregate, such as oolitic or pisolitic hematite,

pisolitic bauxite and oolitic limestone, etc.

(4) Stalactitic aggregates is formed by evaporation of calcitic solution or colloid condensation layer by layer.

(5) Powdery aggregate refers that minerals is dispersed in powder form on other mineral or rock surfaces.

(6) Earthy aggregate refers to the aggregate that is formed by the minerals of loose powder, such as kaolin, montmorillonite, etc.

(7) Filmy aggregate refers that minerals are coated in thin layers on other minerals or rock surfaces, such as malachite.

(8) Dendritic aggregate refers that monomers grow rapidly in some directions like a tree by macle or parallel connection laws.

(9) The salt-like aggregate is a membrane composed of soluble salts, such as saltpeter formed on the ground surface in a dry area.

(10) Crustiform aggregate refers to a thick layer of minerals covering other mineral or rock surfaces.

(11) Massive aggregate refers to the dense block whose boundary line of monomer is not visible to the naked eyes, such as massive chalcopyrite.

1.4.3 Color observation of minerals

Color is a reflection of the absorption and reflection of visible light waves at different wavelengths, such as white calcite and quartz, dark green olivine, copper yellow chalcopyrite, brown limonite, rust hematite and so on. Many minerals have special colors, so they can be used as a feature to identify minerals, such as the special green of malachite, the special blue of the azurite and the special bronze color of the bornite.

There are three main ways to describe the color of minerals as follows:

(1) Standard chromatography

Standard chromatogram, namely, red, orange, yellow, green, blue, purple, white, grey, black are used to describe the colors of minerals. If the color of the mineral is slightly different from the standard color, the appropriate adjectives can be added, such as light green, light green, dark green, dark green, etc.

(2) Analogy method

The most common physical colors are used to describe mineral colors, such as orange red (realgar), grass green (epidote), etc.

(3) Binomial nomenclature

Two standard chromatograms are used to describe mineral color, such as yellow green, gray white, blue gray. The main color is behind, and the secondary color is in front. For example, the main color of blue green is green. It's worth noting that the main color is in the back and the secondary color is in front.

1.4.4 Streak observation of minerals

Streak refers to the color of mineral powders, usually the color of the powder when the mineral is scratched on a white unglazed porcelain plate. Streak can eliminate false colors, weaken other colors, and are more stable than the color of the mineral blocks, so it is one of the most important markers for the identification of dark minerals and gold minerals. The streaks can be consistent with their own colors or not. For example, pyrite is copper yellow and its streak is green black. Hematite may be red, steel gray, ironish black or other colors, but its streak is always cerise. Galena is lead gray and its streak is black.

The method of describing the streak color is the same as the method describing the color of the minerals. The following items should be noted when observing the streaks.

(1) The sharp edge of mineral is used to scratch the whole white porcelain plate (or the bottom edge of the coarse porcelain bowl) lightly, not hard. The streaks of leucocratic minerals are usually colorless, white or light color, but the dark minerals have obvious streaks.

(2) If the streaks do not exist, a knife can be used to scrape the powder which can be observed on a porcelain plate or a white paper, or a small hammer can be used to knock down a small piece and grind into powder. The finer the mineral powder is, the higher the color accuracy of the streaks is.

(3) The finer the mineral powder is, the more obvious the color is, which can help to identify the minerals. For example, both the streaks of graphite and molybdenum are black or ash black, the streak of graphite is still black after grinding, but the streak of molybdenum is green yellow after grinding.

1.4.5 Gloss observation of minerals

Gloss refers to the reflection ability of mineral's smooth surfaces. According to the intensity of reflected light, gloss is divided into two types, namely metallic luster and non-metallic luster. Nonmetallic luster can be subdivided into submetallic luster, adamantine luster and glassy luster.

(1) Metallic luster shows clear metallic light. It is opaque and has black streak, such as natural gold, galena, magnetite, pyrite, chalcopyrite, graphite and so on.

(2) Submetallic luster shows weak metallic light. It is translucent and mainly has dark color streak, such as black sphalerite, hematite and so on.

(3) Adamantine luster shows diamond-like light. It is translucent or transparent and has light color, colorless, and white streaks, such as diamond, light colored sphalerite and scheelite.

(4) Glassy luster shows glassy light. It is transparent and has colorless and white streaks, such as quartz.

The above four types of gloss appear when minerals have flat surface. If the minerals have uneven surfaces or small pores, or are aggregations, they will show some special gloss due to the reflected light on the surface.

(1) Oily luster often appears on the uneven fractures surface of the transparent minerals, such as quartz and talc.

(2) Nacreous luster often appears on the perfect cleavage surface of the transparent minerals, such as mica and iceland spar.

(3) Pitchy luster often appears on the uneven fracture surface of the translucent or opaque black minerals, such as cassiterite, magnetite, pitchblende and so on.

(4) Earthy luster often appears on the surface of powder or soil aggregate minerals, whose surface is dull, such as kaolinite, monohydrallite and limonite.

(5) Silky luster often appears on the fibrous or finely scaled transparent minerals due to the mutual interference of light reflection, such as asbestos and sericite.

When observing the gloss of minerals, different minerals should be compared under the same light intensity degree and rotated slowly from different angles.

The special gloss of minerals is caused by certain special factors, not every mineral is necessary, but every mineral should belong to a certain kind of gloss.

1.4.6 Transparency observation of minerals

Transparency is used to describe the visible light degree passing through the mineral, which depends on the chemical properties and crystal structure of minerals, also affected by the thickness of the minerals. Transparency of mineral samples should be identified with the same thickness, namely, a slice with 0.03mm thickness is usually used.

Depending on the transparent degree of minerals, transparency can be divided into three types, namely transparent minerals, semi-transparent minerals and opaque minerals.

(1) Transparent minerals

Most light can pass through these minerals. The object on the opposite direction can be clearly seen through the thin slice of the mineral, such as colorless crystal and Iceland spar.

(2) Semi-transparent minerals

Only part of light can pass through these minerals. The object on the opposite direction cannot be seen through the thin slice of the mineral clearly, such as talc, cinnabar, blende and opal.

(3) Opaque minerals

Light cannot pass through these minerals, such as pyrite, magnetite and graphite.

It is difficult to identify the transparency of minerals with naked eyes, so the transparency of mineral needs to be judged by its streak.

For opaque minerals, light rarely passes through the powder of mineral, so its streak is often black.

For transparent minerals, much light can pass through the powder of minerals, so its streak is often colorless or white.

For semi-transparent minerals, they can show different absorption degree of light because they are powdery, its streak can present various color, such as red, brown and so on.

1.4.7 Cleavage and fracture observation of minerals

Cleavage and fracture is another important identification character, which affects the mechanical strength of the rocks. Cleavage surfaces refer to the smooth surfaces which are formed when crystalline minerals split along a certain direction under the hitting action. If it is an uneven surface, it is termed as fracture. Cleavage can be divided into four types, namely, perfect cleavage, good cleavage, fair cleavage and poor cleavage.

Perfect cleavage is easy to crack. Cleavage surface is large and smooth. There is no fracture in these minerals, such as mica, graphite, molybdenum and gypsum.

Good cleavage surface is smooth, whose minerals are easily split into laminated or small pieces, such as calcite, galena and fluorite.

Fair cleavage surface is not very smooth, such as amphibole, common pyroxene and scheelite.

Poor cleavage surface may not exit; it often appears fracture, such as quartz and apatite.

The typical features of the cleavage surface are summarized as shown in Table 1-1.

Typical features of cleavage **Table 1-1**

<table>
<tr><th>Cleavage degree</th><th colspan="2">Are there cleavage surfaces easily?</th><th>Smoothness of the cleavage surface</th><th>Development degree of the fracture</th></tr>
<tr><td>Perfect</td><td rowspan="2">Yes</td><td>It is liable to flake</td><td>Perfect</td><td rowspan="4">Poor
↓
Perfect</td></tr>
<tr><td>Good</td><td>It is only liable to cleavage block not flake</td><td>Good</td></tr>
<tr><td>Fair</td><td colspan="2">No</td><td>Fair</td></tr>
<tr><td>Poor</td><td colspan="2">No</td><td>Poor</td></tr>
</table>

It should be distinguished the differences between the cleavage surface and the crystal surface in the process of observation as shown in Table 1-2.

Differences between the cleavage surface and the crystal surface **Table 1-2**

Cleavage surface	Crystal surface
The cleavage surface is along the weak connection direction of the internal crystals, which are continuously parallel to each other under the action of external forces	The crystal surface is a plane on the outside of the crystal, and it disappears immediately after broken
The cleavage surface is generally brighter	The crystal surface is generally dim
The cleavage surface is relatively flat, but there may beregular stepped cleavage or cleavage lines	The crystal surface is not very smooth and often shows uneven marks or various grain pattern when careful observation

There is no fracture when there is perfect cleavage and vice versa. For example, quartz only has fracture and no cleavage surface. There are four main types of fractures of minerals as follows.

(1) Conchoidal fracture

The fracture is a circular smooth surface with irregular concentric stripes, whose shape is like a shell, such as quartz.

(2) Ragged fracture

The fracture is sharply serrated. The malleable minerals have such fractures, such as native copper.

(3) Uneven fracture

The fracture surface is uneven and coarse. Most minerals have such fractures, such as apatite.

(4) Earthy fracture

The fracture surface is fine powder. It is a characteristic rough fracture of earthy minerals, such as kaolinite.

1.4.8 Hardness observation of minerals

Hardness is the most important feature of mineral identification, which refers to the ability to fight curving or grinding by external force. Generally, the relative hardness of minerals is determined by curving each other. The standard of hardness is made up of 10 kinds of minerals, which is called Mohs Hardness Scale as shown in Table 1-3.

Mohs Hardness Scale **Table 1-3**

Scale	1	2	3	4	5	6	7	8	9	10
Minerals	Talc	Gypsum	Calcite	Fluorite	Apatite	Orthoclase	Quartz	Topaz	Corundum	Diamond

It's worth noting that Mohs Hardness Scale only reflects the relative hardness orders of the minerals, not the absolute hardness order.

When hardness of mineral is identified in the wild, a more simple identification method is usually used instead of Mohs Hardness Scale, namely, nails (its Mohs hardness is 2.5) and knives (its Mohs hardness is 5.5) are used to distinguish the hardness of minerals. The hardness of minerals is generally roughly divided into three levels.

The Mohs hardness of common rock forming minerals is mainly between 2 and 6.5. Only a few mineral's Mohs hardness is greater than 6.5, such as quartz, olivine, garnet and so on. The Mohs hardness of fingernail is 2.5, and penknife is 5.5, which are always be used to identify the hardness of minerals. The minerals that can be carved by the nail belongs to low hardness minerals, cannot be carved by the fingernails and can be carved by the knife belongs to medium hardness minerals, cannot be carved by the knife belongs to high hardness minerals.

1.4.9 Specific gravity observation of minerals

Specific gravity of minerals refers to the quality of the mineral over the water quality at 4℃, which must have the same volume.

The specific gravity of minerals is closely related to mineral composition and crystal structure. Minerals with the same ingredients but different structures have different specific gravity. For example, the composition of diamond and graphite is carbon, the specific gravity of diamond is 3.47~3.65, while the specific gravity of graphite is 2.09~2.23.

In generally, most of the minerals with larger specific gravity are metallic minerals. The specific gravity of minerals can be divided into three levels as follows.

(1) Light level

Its specific gravity is less than 2.5. For example, the specific gravity of graphite is 2.2, the natural sulfur is 2.05~2.08, halite is 2.1 ~ 2.2 and gypsum is 2.3.

(2) Middle level

Its specific gravity is between 2.5 and 4.0. For example, the specific gravity of orthoclase is 2.55~2.75, the plagioclase is 2.61~2.76, fluorite is 3.18 and diamond is 3.5.

(3) Heavy level

Its specific gravity is more than 4.0. For example, the specific gravity of the barite is 4.3~4.7, magnetite is 4.6~5.2, scheelite is 5.8~6.2, galena is 7.4~7.6 and natural gold is 15.6~19.3.

1.4.10 Magnetism observation of minerals

The magnetism of minerals refers that minerals can be attracted by permanent magnets or electromagnet, or that minerals themselves can attract the iron objects, mainly because they contain elements such as iron, cobalt, nickel, chromium, titanium and vanadium. Minerals can be divided into four kinds according to their magnetism.

(1) Strongly magnetic mineral can be attracted by ordinary horseshoe magnet, such as magnetite.

(2) Medium magnetic mineral cannot be attracted by ordinary horseshoe magnet, but can be attracted by weak electric magnet, such as ilmenite.

(3) Weakly magnetic minerals only can be attracted by strong electric magnet, such as monazite.

(4) Nonmagnetic minerals even cannot be attracted by strong electric magnet, such as corundum.

1.4.11 Luminescent observations of minerals

When some minerals are stimulated by external energy, such as ultraviolet, x-rays, cathode rays, radioactive rays, or shock, friction and heat, they emit visible light. If there is no excitation energy from outside, the minerals continue to light, which is called phosphorescence. If there is no excitation energy from outside, the minerals cannot light immediately, which is called fluorescence. An object that can fluoresce or phosphor is called a fluorophor or phosphor.

Luminescence can be used to identify some minerals including diamond, scheelite, zircon, fluorite and so on.

1.4.12 Sensory characteristic observation of minerals

Some minerals have properties that can be felt by the human senses. For example, talc and graphite has smooth feel, the diatomite has a rough feel, and potash has a bitter taste. All of these properties can be used for mineral identification.

1.4.13 Other characteristics observation of minerals

The special properties of minerals are of great significance for minerals identification, which are described as follows.

(1) Elasticity

Elasticity refers to the property that mineral will bend and deform when it is subjected to external force and it can restore the bending deformation when the external force stops, such as mica and asbestos.

(2) Flexibility

Flexibilty refers to the property that mineral bends and deforms when it is subjected to external forces, and bending deformation of the mineral cannot be restored to its original nature when the external force is removed, such as talc and chlorite.

(3) Fragility

Fragility refers to the property that minerals are easily broken by external forces. For example, even though the hardness of specular hematite is bigger than knife, it still can be crushed into small grains or powders by a knife due to its obvious fragility.

(4) Malleability

Malleability refers to the property that minerals are easy to form thin sheets and filaments under hammer or pull, such as native gold, native silver and native copper.

Montmorillonite has water expansibility and disintegration. Carbonate minerals have hydrochloric acid reaction. For example, the calcite produces strong bubbles when it meets dilute hydrochloric acid, but dolomite only produces weak bubbles when it is cut into powder by knife to meet dilute the hydrochloric acid.

When minerals cannot be identified by the naked eye, simple chemical experiments must be used to identify them. The primary chemical composition or some physical properties of the minerals are tested by simple chemical reagents to identify the minerals with the appearance of minerals or other microscopic data. These chemical methods include bead dyeing reaction, cobalt nitrate experiment, flame test, spot test, phosphoric acid process, powder grinding method, microcrystalline analysis method, staining method and striation method. These experimental methods still have been used today because they are simple and easy, but they can only be used as supplement methods to the naked eyes because of their limitations. Table 1-4 shows the characteristics of the main rock forming minerals.

Characteristics of the main rock forming minerals **Table 1-4**

Minerals	Chemical component	Shape	Color	Streak	Gloss	Transparency	Cleavage and fracture	Hardness	Specific gravity	Chemical properties and others	Distribution	Main identification characteristics	Sample
Quartz	SiO_2	Complete crystal shape is hexagonal prism or bipyramid. But it is mainly granular, sometimes is crystal cluster	Pure quartz is colorless or milk white. When quartz contains impurities, it may show different colors, such as purple red, pink and other colors	None	Quartz shows glassy luster, and its fracture shows oily luster	Transparent	Conchoidal fracture	7	2.65	It is stable and has anti-weathering ability	Quartz can be found in or out of igneous rocks, sedimentary rocks and igneous rocks especially in acidic igneous rocks in the form of monocrystalline crystal clusters and nervation	Shape and hardness	
Orthoclase	$KAlSi_3O_8$	Columnar, clintheriform or granular	Flesh red, light-rose color or almost white	None	Glassy luster	Semitransparent or opaque	Two sets of perfect orthogonal cleavages	6	2.55~2.75	Orthoclase is easily weathered to kaolinite	Orthoclase mainly distributes in granite, syenite gneiss and other igneous rock	Cleavageand colors It is easily weathered	

continued

Minerals	Chemical component	Shape	Color	Streak	Gloss	Transparency	Cleavage and fracture	Hardness	Specific gravity	Chemical properties and others	Distribution	Main identification characteristics	Sample
Anorthose	(Na,Ca)[$AlSi_3O_8$]	Columnar and clintheriform	White or grey-white	White	Glassy luster	Semitransparent or opaque	Two sets of good oblique (86°) cleavages Its fracture is flat	6	2.61~2.76	Anorthose is easily weathered to kaolinite	Those mainly containing Na are only found in acidic or intermediate igneous rocks. Those mainly containing Ca are only found in intermediate igneous rocks or basic igneous rocks	Color There are striolas on the cleavage surface	
White mica	KAl_2[$AlSiO_{10}$]$(OH,F)_2$	Flaky and schistose	It is colorless; sometimes it is gray, light yellow, light red and so on	None	Glassy luster or nacreous luster	Transparent	A set of perfect cleavages	2.5~3	2.76~3.1	It is elastic	It is widely distributed in igneous rocks	Cleavage Its slice is elastic	

continued

Minerals	Chemical component	Shape	Color	Streak	Gloss	Transparency	Cleavage and fracture	Hardness	Specific gravity	Chemical properties and others	Distribution	Main identification characteristics	Sample
Black mica	$K(Mg,Fe)_3(AlSi_3O_{10})(OH,F)_2$	Flaky and schistose	Black and brown-black	None	Nacreous luster	Transparent	A set of perfect cleavages	2.5～3	3.02～3.12	It is elastic. Under hydration, it lost elasticity and became vermiculite	It is widely distributed in igneous rocks	Cleavage and color Its slice is elastic	
Hornblende	$(Ca,Na)_{2-3}(Mg,Fe,Al)_5[Si_6(Si,Al)_2O_{22}](OH,F)_2$	Long columnar or fibrous	Green black	Light green	Glassy luster	Semitransparent or opaque	Crossing angle of two sets cleavages is nearly 56°	5.5～6	3.0～3.4	After being heated by water, it can be turned into chlorite or serpentine	It mainly distributes in intermediate igneous rocks, such as diorite, andesite. It is also possible to form ultrabasic amphibolite alone	Shape and color The crystal cross-section is nearly octagonal	

continued

Minerals	Chemical component	Shape	Color	Streak	Gloss	Transparency	Cleavage and fracture	Hardness	Specific gravity	Chemical properties and others	Distribution	Main identification characteristics	Sample
Pyroxene	(Ca, Mg, Fe, Al) $[(Si,Al)_2O_6]$	Short column The section is octagonal They are often granular in rocks	Dark black, brown black or purple black	Grey green	Glassy luster	Semitransparent or opaque	Two sets of good or fair cleavages are nearly orthogonal	5～6	3.15～3.9	After being heated by water, it can be turned into chlorite or serpentine	It is mainly found in the basic igneous rocks, such as gabbro and basalt. It can also form ultrabasic pyroxenite alone	Shape and color The crystal cross-secti on is nearly octagonal	
Olivine	$(Mg, Fe)_2SiO_4$	It is often granular aggregate	Olive green or light yellow green	None	Glassy luster and oily luster	Semitransparent or opaque	Usually there is no cleavage, only conchoidal fracture	6.5～7	3.27～3.48	When olivine is dissolved in sulfuric acid, SiO_2 gel precipitates out	Olivine only can be found in basic igneous rocks. It also can be made into peridotite.	Color and hardness It is not symbiosis with quartz	

continued

Minerals	Chemical component	Shape	Color	Streak	Gloss	Transparency	Cleavage and fracture	Hardness	Specific gravity	Chemical properties and others	Distribution	Main identification characteristics	Sample
Calcite	$CaCO_3$	Rhomb or granularity	White, grey-white It is brown red when containing Fe, and brown black when containing Mn	None	Glassy luster	Transparent or semitranspare	Three sets of good cleavages	3	2.71	When calcite meets the hydrochloric acid, it will produce strong bubbles	It is widely found in limestone, and also can be found in marble. It can be found in some of the igneous rocks in small amounts. It can also appear as a calcite vein	Cleavage and hardness It can produce bubbles when meeting hydrochloric acid	
Dolomite	$CaCO_3 \cdot MgCO_3$	It is often a rhombohedral block. The crystal surface often can be bent into a saddle	White, light yellow or light red	White	Glassy luster	Transparent or opaque	Three sets of good cleavages	3.5~4	2.8~2.9	When dolomite meets the hydrochloric acid, it will produce a few bubbles. It can be used to distinguish from the calcite	Dolomite mainly distributes in dolomite. Sometimes it can be found in marble or limestone	Cleavage and hardness Its crystal surface can be curved. It can produce bubbles when meeting hydrochlo ric acid	

continued

Minerals	Chemical component	Shape	Color	Streak	Gloss	Transparency	Cleavage and fracture	Hardness	Specific gravity	Chemical properties and others	Distribution	Main identification characteristics	Sample
Gypsum	$CaSO_4 \cdot 2H_2O$	Clintheriform, strip or fibrous aggregate	Colorless, white or gray white	White	Glassy luster The fibrous gypsum shows silky luster	Semitrans-parent or opaque	A set of cleavages are developed	2	2. 3	Gypsum can dissolve in hydrochloric acid and also slightly soluble in water. It has slippery feeling, flexibility and small hardness	Gypsum is the sediment mineral of lagoonal facies or estuarine facies	Cleavage and hardness It is flexible	
Kaolinite	$Al_4[Si_4O_{10}](OH)_8$	Flaky or dense fine granular aggregate	Flake is colorless and block is white	White	Earthy luster	Opaque	A set of good cleavages or earthy fracture	1	2. 16～2. 68	Kaolinite is smooth. It is viscous and malleable after absorbing water	Kaolinite is a clay mineral that is formed by weathering of such as feldspar and augite, and has a wide distribution	It is soft, absorbent and malleable	

continued

Minerals	Chemical component	Shape	Color	Streak	Gloss	Transparency	Cleavage and fracture	Hardness	Specific gravity	Chemical properties and others	Distribution	Main identification characteristics	Sample
Talc	$Mg_3[Si_4O_{10}](OH)_2$	Schistose or massive	White, light red or light grey	White or green	Oily luster or nacreous luster	Semitransparent or opaque	A set of good cleavages	1	2.6～2.8	Talc is so smooth. It is soft and its slice is flexible	Talc is the main minerals formed after metamorphism of olivine, pyroxenite and hornblende	Color and hardness It is very smooth	
Chlorite	$(Mg,Al,Fe)_{12}[(Si,Al)_8O_{20}](OH)_2$	Schistose, clintheriform or small scaly aggregate	Dark green	None	Nacreous luster	Semitransparent oropaque	A set of good cleavages	2～2.5	2.6～3.3	Scales and flakes are inelastic and flexible	Chlorite has a wide distribution in igneous rocks and often constitutes chlorite schist	Color and slice It has no flexibility and has flexure	

continued

Minerals	Chemical component	Shape	Color	Streak	Gloss	Transparency	Cleavage and fracture	Hardness	Specific gravity	Chemical properties and others	Distribution	Main identification characteristics	Sample
Serpentine	$Mg_6[Si_4O_{10}](OH)_8$	Compact massive, schistose or fibrous	Light green or dark green	White	The lump is waxy luster. The fibrous gypsum shows silky luster	Semitransparent or opaque	uneven fracture	2.5~3.5	2.44~2.8	Serpentine can be gotten by the metasomatic reaction of olivine and pyroxene. It can dissolve in hydrochloric acid	Serpentine often distributes with asbestos. Most of them are igneous minerals of ultrabasic rocks	Color and glossy It is smooth	
Garnet	$(Ca,Mg)(Al,Fe)[SiO_4]_3$	Rhombic dodecahedron or icositetrahedron The aggregate is granular	Brown, brown red or green dark	Colorless	Glassy luster Its fracture is oily luster	Opaque	There is not cleavage. The fracture is not smooth	6.5~7.5	3.32~4.19	It is more stable. It can be changed into the limonite and so on	Garnet is standardi gneous mineral and can be found in igneous rock	Shape, color and hardness	
Pyrite	FeS_2	Cube, massive or concretion form	Bassy yellow	Dark green	Metallic luster	Opaque	Conchoidal fracture or irregular fracture	6~6.5	4.9	Under the action of oxidation or water, pyrite can form limonite. Its crystal surface has streaks	Pyrite often can be found in sandstone and limestone	Shape, color and gloss There are streaks on its crystal surface	

1. 5 Internship report

Please observe the characteristics of samples and fill the observation results in Table 1-5.

Identification of unknown minerals **Table 1-5**

Number	Color	Shape	Gloss	Hardness	Cleavage	Fracture	Other	Mineral name

1. 6 Thinking questions

(1) Please list the identification characteristics and main differences of the rock minerals in the following groups.

① Orthoclase, anorthose and quartz

② Hornblende, pyroxene and black mica

③ Calcite, dolomiteand and gypsum

(2) What are the main differences among quartz, feldspar and calcite?

(3) What are the main characteristics of minerals identified by the naked eyes?

Chapter 2 Rock identification

2.1 Purposes and requirements

(1) To understand the form cause and classification of three kinds of rocks, master the basic characteristics of igneous rocks, sedimentary rocks and metamorphic rocks by observation and identification of typical rock samples.

(2) To master the composition, structure and texture features of main rocks in civil engineering, preliminarily understand its engineering properties by identification of rock samples.

2.2 Common tools

Common tools include magnifying glass, knifes and diluted hydrochloric acid.

2.3 Identification method

The minerals composition, structure and texture of rocks are different due to the different forming environment. The first step of identification is to judge the type of the rock sample according to its color, namely sedimentary rock, igneous rock or metamorphic rock. The second step is to determine the name of the rock sample according to its main mineral components and their contents, then to comprehensively analyze the rock sample combined with its structure and texture characteristics.

2.3.1 Texture differences of three kinds of rocks

There are obvious texture differences of the three kinds of rocks as follows.

(1) Igneous rock

The igneous rock has obvious crystalline texture because it is formed directly from the molten magma of high temperature. The degree of crystallinity is one of the most important items of texture. An igneous rock may be composed of an aggregate of crystals, of natural glass, or of crystals and glass in varying proportions, which depends on the rate of cooling and composition of the magma on the one hand and the environment under which the rock developed on the other. If a rock is completely composed of crystalline mineral material, it is described as holocrystalline. Most rocks are holocrystalline. Conversely, rocks that consist entirely of glassy material are referred to as holohyaline. The terms hypo-, hemi- or merocrystalline are given to rocks that are made up of intermediate proportions of crystalline and glassy materials.

The crystallization texture of igneous rocks is shown in the sequentiality of condensation crystallization successively.

(2) Sedimentary rock

The sedimentary rock is formed by the weathering loose materials of the pre-exist rocks, which is transported, deposited, compacted and consolidated, and has obvious sedimentary environment characteristics. Sedimentary characteristics mainly focus on the regularity, size and shape of the composition particles and its combinational relationship, which can present clastic texture, such as rudaceous texture, sand texture, silty texture, pelitic texture and biochemical crystal texture and other phenomena.

The crystalline texture of sedimentary rocks is reflected in the chemical properties of the mineral compositions by precipitation or recrystallization in the solution.

(3) Metamorphic rock

Different metamorphic rocks are formed by different pre-exist rocks due to different degrees of metamorphic factors. The metamorphic degree is influenced by metamorphic factors and the pre-exist rock. Metamorphic rock has the inheritance and uniqueness, which presents the crystalloblastic texture, palimpsest texture and cataclastic texture.

The crystalloblastic texture of metamorphic rocks is characterized by the orientation of various minerals in solid state due to simultaneous recrystallization.

2.3.2 Structure differences of three kinds of rocks

(1) Igneous rock

Igneous rocks present different structure phenomena affected by the magma property, occurrence condition, space motion state of the lava flow and so on. Intrusive rocks have massive structure, and extrusive rocks have rhyotaxitic structure, vesicular structure or amygdaloidal structure.

(2) Sedimentary rock

There are differences in lithology of sedimentary rocks and sedimentary facies affected by the external force property, paleogeographic environment, material origin and depositional condition. However, both sedimentary rocks and sedimentary facies show obvious stratified structure features, such as bedding and lamellation. Sedimentary rocks also have chap, ripples and other hypergene structures and paleontological fossils. All of these features are used to distinguish sedimentary rocks from igneous rocks and metamorphic rocks.

(3) Metamorphic rock

The structures of metamorphic rocks are mainly composed of schistosity structure and gneissic structure which are formed due to directional arrangement of minerals, because the metamorphic environment, mode and degree of the pre-exist rocks are different.

2.4 Identification of igneous rocks

2.4.1 Classification of igneous rocks

There are many kinds of igneous rocks in nature, which can be divided into extrusive rocks and intrusive rocks according to the geological environment of igneous rock. It can be divided into four basic types according to the SiO_2 content, namely ultrabasic rocks (the content of SiO_2 is less than 45%), basic rock (the content of SiO_2 is between 45% and 52%), intermediate rock (The content of SiO_2 is between 52% and 65%) and acid rock (the content SiO_2 is more than 65%) as shown in Table 2-1.

Classification of igneous rocks **Table 2-1**

<table>
<tr><td colspan="4">Type of igneous rocks</td><td>Acid rock</td><td>Intermediate rock</td><td>Basic rock</td><td>Ultrabasic rock</td></tr>
<tr><td colspan="4">Chemical components</td><td colspan="2">Rich in Si and Al</td><td colspan="2">Rich in Fe and Mg</td></tr>
<tr><td colspan="4">Content of SiO_2</td><td>>65%</td><td>52%~65%</td><td>45%~52%</td><td><45%</td></tr>
<tr><td colspan="4">Color</td><td colspan="4">Light(light grey, yellow, brown and red)→dark (dark grey, black green and black)</td></tr>
<tr><td colspan="4">Features</td><td colspan="2">Orthoclase</td><td>Anorthose</td><td>Excluding feldspar</td></tr>
<tr><td colspan="4">Main minerals
Occurrence / Structur / Texture</td><td>Quartz, black mica and hornblende</td><td>Black mica and hornblende</td><td>Pyroxene, hornblende and olivine</td><td>Olivine, pyroxene and hornblende</td></tr>
<tr><td rowspan="2">Extrusive rocks</td><td rowspan="2">Volcanic vent, volcanic cone, lava flow and lava sheet</td><td>Massive structure, rhyotaxitic structure, vesicular structure and amygdaloidal structure</td><td>Cryptocrystal, vitric or colloidal state</td><td>Rhyolite</td><td>Trachyte</td><td>Basalt</td><td>Nephelinite</td></tr>
<tr><td>Massive structure and vesicular structure</td><td>Colloidal state</td><td colspan="2">Pumice and obsidian</td><td colspan="2">It is rare</td></tr>
</table>

continued

Type of igneous rocks					Acid rock	Intermediate rock	Basic rock	Ultrabasic rock
Intrusive rocks	Hypabyssal rock	Laccolite and dyke	Massive structure and vesicular structure	Equigranular state, porphyaceous state or maculosus state	Granite porphyry	Orthophyre	Diabase	Kimberlite
	Plutonite	Batholith	Massive structure	Granulous state, pleocrystalline	Granite	Syenite	Gabbro	Olivinite and pyrozenite

2.4.2 Color of igneous rock

The content of dark colored minerals plays a decisive role in the color of igneous rocks. When the igneous rocks mainly contain dark colored minerals and present dark color, they are usually ultrabasic or basic rocks. When the igneous rocks contain few dark colored minerals and present light color, they are usually acid or intermediate rocks. In addition, the color of the rock is related to the crystallization degree of the rock. In general, the color of the rock with cryptocrystalline texture is darker than the coarse-grained crystalline rock with the same composition.

The color ratio method is often adopted to observe the rock's color, namely, to estimate the area percentage of dark minerals in light minerals on the surface of the rock that are visible. The igneous rocks can be divided into 4 types according to the content of dark minerals as follows.

(1) Leucocratic rock

The color ratio of the rocks is between 0 and 35%, which belongs to the acid rock.

(2) Mesocratic rock

The color ratio of the rocks is between 35% and 65%, which belongs to the intermediate rock.

(3) Melanocratic rock

The color ratio of the rocks is between 65% and 90%, which belongs to the basic rock.

(4) Dark rock

The color ratio of the rocks is between 90% and 100%, which belongs to the ultrabasic rocks.

The color of the rocks can be simplified when they are estimated by naked eyes, namely, the rocks with about a third of dark minerals on the surface are leucocratic rocks, with 2/3 of dark minerals are mesocratic rocks, with more than 2/3 of dark minerals are melanocratic rocks. When there are only few light minerals in dark minerals, it is dark rocks.

It's worth noting that, not only the overall color of the samples should be observed, but also the color of fresh rock or fresh surface should be paid attention when looking at the color of the rock.

2.4.3 Mineral composition of igneous rock

According to the identification methods of minerals, the minerals composition and their combination style, the combination characteristics, the main minerals and secondary minerals need to be estimated.

For example, the content of the main mineral quartz in granite is about 25%, the content of feldspar is about 60%, and the total content of secondary minerals black mica and amphibole is about 5%. The main mineral of pyroxenite is pyroxene, which is about 90% to 100%, and the secondary minerals include olivine, hornblende, black mica, chromite, magnetite, ilmenite and so on.

The following items should be paid attention when observing the minerals of igneous rocks.

(1) Quartz

Quartz in granite and rhyolite rock mainly shows granular, oily luster and smoke gray in color. Its Mohos hardness is 7. It is easily confused with the gray white anorthose.

(2) Feldspar

Feldspar in granite, diorite and andesite rocks mainly shows glassy luster. Its Mohs hardness is 6. Orthoclase is mostly flesh red and anorthose is mostly gray white. When you observe carefully, you can find many parallel crystal lines in anorthose. The fresh cleavage surface of the orthoclase usually shows two parts with significant light difference seen in the light.

(3) Mica

Mica flake in biotite granite can be peeled with a knife easily.

(4) Pyroxene and hornblende

Gabbro and diorite, pyroxene and hornblende in igneous rocks are both dark gray and black. They have similar luster but different shapes and fracture surfaces.

The crystalline form of pyroxene is short columnar and its cross-section is octagonal and approximately square. Pyroxene is often associated with olivine and has two sets of nearly orthogonal fair cleavages.

The crystalline form of hornblende is long columnar and the cross-section is hexagon. Hornblende is often associated with black mica and has two sets of oblique fair cleavages with angle of 124° or 56°.

2.4.4 Texture and structure of igneous rock

The texture and structure features of igneous rocks are very important characteristics to differentiate from sedimentary rocks and metamorphic rocks, and also the basis for analyzing the formation environment of igneous rocks.

Extrusive rock and intrusive rock can be distinguished according to their texture and structure. Generally, the intrusive rock has holocrystalline gravel texture and massive structure such as granite. Most of the extrusive rocks have cryptocrystalline or vitreous texture and vesicular structure such as basalt.

However, special attention should be paid to the igneous rocks with similar structures and different formation causes. For example, rocks with fine grained structure can be found in the extrusive rocks and may be found at the edge of the intrusive rock. Therefore, it is necessary to consider the field occurrence and distribution laws of rocks when identifying igneous rocks.

2.4.5 To name igneous rock

The following items need to be considered when to name an igneous rock.

(1) The igneous rock is named according to its main mineral with the most content in the rock. The main mineral refers to the mineral with a content greater than 20% in the rock, and the mineral that can determine the division category and naming of the igneous rock.

A rock composed of pyroxene and anorthose can be named as gabbro; a rock composed of hornblende and anorthose can be named as diorite; and a rock composed of feldspar and quartz can be named as granite. The rock that mainly contains orthoclase can be named as orthophyre. The rock that mainly contains anorthose and hornblende can be named as dioritic porphyrite.

(2) Besides the main mineral in the igneous rocks, there are less secondary minerals with less content, which can be used to supplement the categories name of igneous rocks. Secondary mineral is mineral that account for about 3% to 20% of the rock, which does not play a major role in classification and nomenclature but can be used as a mineral to determine the basis of the category.

When the diorite contains a certain amount of quartz, although it does not affect the naming, it should be called quartz diorite, and secondary mineral should be placed before the name of the rock.

(3) There are some other naming methods besides the method that rocks can be named according to the mineral composition. For example, trachyte is named after its macroscopic features; andesite (the Andes) and kimberlite (place name in South Africa) are named after their birthplaces, dioritic porphyrite is named after its texture; rhyolite

and pumice are named after their structures.

2.4.6 The characteristics of common igneous rock

2.4.6.1 Acid rocks

(1) Granite

Granite is one of acid plutonites. It is the main component of continental crust and distributes widely. The main constituent minerals of granite are feldspar (its content is about 60%) and quartz (its content is more than 25%) followed by black mica, occasionally white mica and amphibole. Granite is flesh red, gray white or gray. Granite has massive structure and isometric granular texture, namely holocrystalline medium coarse grain texture.

Granite can be used as material for advanced building decoration engineering, hall floor and outdoor carving engineering, because it is hard and has anti-weathering ability and beautiful color.

(2) Granite porphyry

Granite porphyry is one of hypabyssal rocks and has similar constituent with granite. Granite porphyry has porphyritic texture, whose phenocryst is feldspar or quartz and whose ground mass consists of fine feldspar, quartz and other minerals. Granite porphyry has massive structure. It is grey white or fresh red.

(3) Rhyolite

Rhyolite is a widely distributed extrusive rock and has similar constituent with granite. Rhyolite has porphyritic texture, whose tiny phenocryst is often made up of feldspar or quartz and whose ground mass is made up of cryptocrystalline material or glassy material. The quartz phenocryst in rhyolite is usually smoky gray and round, and has oily luster fracture. The orthoclase phenocryst in rhyolite has approximately square crystalline form and clear glass luster. Rhyolite has rhyotaxitic structure or massive structure. Rhyolite shows different colors such as light red, light yellow, grey white, purple gray and light fawn. Dark minerals such as black mica and amphibole rarely appear in rhyolite.

2.4.6.2 Intermediate rock

(1) Syenite

Syenite is plutonite. Its main minerals include orthoclase and anorthose (more than 60% of the total content)with a small amount of pyroxene, black mica and so on, followed by hornblende (about 20%). Syenite has pleocrystalline equigranular texture and massive structure. Syenite is flesh red, light gray or light yellow.

The physical and mechanical properties of syenite are similar to that of granite, but not as hard as granite and easy to be weathered. The difference from granite is that syenite contains little or no quartz.

(2) Orthophyre

Orthophyre is hypabyssal rock and has same composition minerals with syenite. It has massive structure. Orthophyre has porphyritic texture, whose phenocryst is mainly orthoclase (orthoclase is a good clintheriform crystal which has glassy luster and clear cleavages) followed by anorthose, hornblende and black mica, whose ground mass is solid fine grained cryptocrystalline. Orthophyre is grey white, taupe or light red brown.

(3) Trachyte

Trachyte is extrusive rock. Trachyte has porphyritic texture, whose phenocryst is mainly orthoclase (long strips orthoclase microcrystals are almost parallel), and whose ground mass is cryptocrystalline or vitric. Trachyte has fine pores and rough surface. Trachyte has massive structure, vesicular structure or amygdaloidal structure. Trachyte is usually light grey or light red.

(4) Diorite

Diorite is plutonite. The main minerals of diorite are anorthose (its content is more than 50%, with white or gray short columnar) and hornblende (it accounts for about a third of the total mineral, with dark black green, long column or needle) followed by black mica and pyroxene. Diorite has pleocrystalline equigranular texture and massive structure. Diorite shows different colors such as grey white, grey, grey green and dark gery. Diorite is one of good building stones because it has dense structure, high strength and high anti-weathering ability.

(5) Dioritic porphyrite

Dioritic porphyrite is hypabyssal rock. Dioritic porphyrite has porphyritic texture, whose phenocryst is mainly anorthose and hornblende, occasionally black mica, and whose ground mass is fine grained or cryptocrystalline anorthose. Dioritic porphyrite has massive structure and shows grey or grey green in color.

(6) Andesite

Andesite is extrusive rock. Andesite has porphyritic texture, whose phenocryst is mainly anorthose and whose ground mass is cryptocrystalline or vitric. Andesite has massive structure, vesicular structure or amygdaloidal structure and shows different colors such as grey, purple, grey purple, red brown and light yellow.

2.4.6.3 Basic rocks

(1) Gabbro

Gabbro is plutonite. Its main minerals are anorthose (it is gray or dark gray long strip clintheriform crystal) and pyroxene (it is black, dark green or dark brown short columnar) followed by olivine, hornblende and black mica. Gabbro has pleocrystalline medium-big equigranular texture and massive structure. Gabbro shows grey black, black or black green in colors. Gabbro is one of good road building materials because it has high strength and good anti-weathering ability.

(2) Diabase

Diabase is one of hypabyssal rocks and has similar constituent with gabbro. Diabase is composed of tiny long strip anorthose phenocryst and the granulous pyroxene in the lacuna of anorthose. Diabase has fine grained texture or cryptocrystalline dense texture. It mainly has massive structure, sometimes has vesicular structure or amygdaloidal structure. Diabase is dark light, black grey green or dark grey in color.

(3) Basalt

Basalt is the most widely distributed rock in extrusive rocks. It has similar composition with gabbro. Anorthose in basalt is in the form of long and thin strip, and pyroxene is granulous. Basalt has cryptocrystalline texture or fine grained texture, occasionally porphyritic texture. Basalt has massive structure, vesicular structure or amygdaloidal structure. Basalt shows different colors such as grey dark, dark, dark grey, red brown or dark green. Basalt is hard and brittle. It has extremely developed columnar fissures in the wild, whose vertical columnar fracture section is often like the back of turtle.

2.4.6.4 Ultrabasic rocks

(1) Olivinite

The main mineral of olivinite is olivine, which accounts for more than 10%, and pyroxene. It is rich in minerals such as iron and magnesium. Olivinite is the most common rock type in ultrabasic rocks. The fresh olivinite is olive green and has large specific gravity. It can be weathered into the soil in the moist and warm environment. Natural diamond is mainly produced in kimberlite, and kimberlite is made from olivinite, so olivinite is the basic source of natural diamond.

(2) Kimberlite

The old name of the kimberlite is breccia mica olivinite. In 1887, it was discovered in Kimberley, South Africa, and then it named after the name of this place.

Kimberlite is often black, dark green or gray in color. It has porphyritic texture. It is one of the main igneous rocks that produce diamond.

(3) Pyroxenite

Its main mineral is pyroxene, which accounts for 90% to 100%. Pyroxenite usually contains a small amount of other minerals such as olivine, hornblende, black mica, chromite, magnetite, ilmenite and so on. The pyroxenite is often dark and has granular texture. It can be eroded into fibrous serpentine easily.

2.4.7 Identification method of igneous rocks by naked eyes

The main basis for the identification of igneous rocks is the occurrence in the field, texture, structure, mineral composition and color, etc. The identification procedure is as follows.

(1) Plutonite, hypabyssal rock and extrusive rock should be differentiated in combination with their occurrence in the field, texture and structure. Their identification characteristics are shown in Table 2-2.

Characteristics of igneous rocks **Table 2-2**

Characteristics	Plutonite	Hypabyssal rock	Extrusive rock
Occurrence	Most plutonites appear as large intrusive bodies such as batholiths. Some plutonites appear as the rock basin and laccolith. There is obvious metamorphic belt in the surrounding rocks near the contact zone	Hypabyssal rock mainly appears as dyke or dike rock wall. There may be a narrow contact metamorphic belt in the surrounding rocks	Extrusive rock appears as lamellar or irregular stratiform, volcanic cone, lava flow or spread of lava. There is generally not metamorphic belt in the surrounding rock
Structure	Plutonite often has massive structure	Hypabyssal rock often has massive structure, sometimes has a little small pores. Generally, there is no amygdaloid structure	Extrusive rock often has vesicular structure, amygdaloid structure or rhyotaxitic structure
Texture	Plutonite often has epigranular (medium grain or coarse grain) pleocrystalline texture. Porphyritic like texture may appear in the rock mass center	Hypabyssal rock has fine grained texture or porphyritic texture, whose ground mass is mainly fine grain or cryptocrystalline	Extrusive rock has porphyritic texture, cryptocrystalline texture or vitreous texture

(2) Observing the color of the igneous rocks

The color of the igneous rock reflects its chemical and minerals composition to a large extent. According to the SiO_2 content in the chemical composition, igneous rocks are classified into ultrabasic, basic, intermediate and acid rocks. The specific content of SiO_2 is indistinguishable to the naked eyes, but its content is often reflected in the mineral composition. In general, the content of SiO_2 is high, the light color mineral is more, the content of SiO_2 is low, and the dark color mineral is relatively more.

The mineral color is the dominant factor that makes up the color of rock, so the color can be used as one of the characteristics to distinguish the igneous rocks. From ultrabasic rock to acid rock, the color goes from dark to light. In general, ultrabasic rocks are black, black green and dark green, and the basic rocks are gray black and gray green. The intermediate rocks are gray and gray white, and the acid rocks are fresh red, light red and white.

(3) Observing mineral composition

The first observation is whether there is quartz. When there is quartz, the quantity of quartz should be observed. The second observation is whether there is feldspar, and when there is feldspar, we should try to distinguish whether it is orthoclase or anorthose. Moreover, it is important to note the black mica, which is often found in acid rocks.

(4) The key identification points of igneous rocks are described below.

① Plutonite

Plutonite is often characterized by an equigranular pleocrystalline texture, whose mineral particles are coarse and in direct contact with each other without cementation. It is easier to identify the plutonite.

② Hypabyssal rock

When phenocryst exits in hypabyssal rock, hypabyssal rock can be divided into two main types according to the mineral composition of the light color porphyry minerals. It is called porphyrite when phenocryst is mainly anorthose, and it is called porphyry when phenocryst is mainly orthoclase or quartz.

If the porphyrite also contains amphibolites phenocryst or hornblende ground mass, it is called dioritic porphyrite. If the phenocryst only contains orthoclase without quartz, porphyrite is called orthophyre. If the phenocryst contains not only quartz but also orthoclase, porphyrite is called granite porphyry. If the phenocryst only contains quartz, porphyrite is called quartz porphyry.

For the fine grain texture hypabyssal rock, it can be named according to the classification table of igneous rocks when the mineral composition and color can be determined. It should be added fine-grained before its name, such as fine-grained granite.

For the vein rock with cryptocrystalline texture, it can be named as "light color vein rock" (such as felsites) and "dark color vein rock" (diabase) based on its color.

③ Extrusive rock

It is difficult to identify extrusive rocks by naked eyes. The ground mass is usually fine grained or vitreous texture except phenocryst. Extrusive rocks only can be named preliminarily based on the overall consideration of color, phenocryst composition, texture and structure, etc. The characteristics of the common extrusive rocks are shown in Table 2-3.

Identification characteristics of several extrusive rocks by naked eyes **Table 2-3**

Characteristics	Basalt	Andesite	Trachyte	Rhyolite
Color (fresh rock)	It is black green or black. Its glossy is dark	It is grey red, grey purple or brick red	It is light grey, light red or grey purple	It is pink, grey green or light grey purple
Phenocryst composition	Its minerals include pyroxene, anorthose and olivine	Its minerals mainly include anorthose, sometimes pyroxene, hornblende and black mica	Its minerals include potassium feldspar, black mica and hornblende	Its minerals include quartz which often appears corrosion phenomenon and potassium feldspar
Texture	It often has compact texture, fine grained texture or cryptocrystalline texture	It often has porphyritic texture or cryptocrystalline texture	It often has porphyritic texture or cryptocrystalline texture	It often has porphyritic texture, cryptocrystalline texture or vitreous texture
Structure	It has vesicular structure and amygdaloidal structure	Sometimes, it has vesicular structure and amygdaloidal structure	It has massive structure, sometimes it has vesicular structure	It has rhyotaxitic structure, vesicular structure and amygdaloidal structure

2.4.8 An identification example of igneous rock

Example: The rock is fresh red, whose weathering surface is yellow. It has massive structure and medium grained texture. Its main minerals are quartz, orthoclase, anorthose and a small amount of black mica described as follows.

(1) Quartz is colorless and has oily luster fracture. It accounts for 25% of the total content.

(2) Orthoclase is fresh red and tabular. It has perfect cleavage and glassy luster. It accounts for 55% of the total content.

(3) Anorthose is grey white and tabular. It has perfect cleavage and glassy luster. It accounts for 15% of the total content.

(4) Black mica is black and schistose. It has perfect cleavage and nacreous luster. It can be cut off the small pieces by a knife. It accounts for 3% of the total content.

(5) It contains magnetite and other accessory minerals.

Identification of igneous rock

The rock can be named after granite according to its main minerals including a lot of quartz and feldspar. It can be named after biotite granite further because its secondary mineral is black mica. At last the rock can be named after fresh red medium grain biotite granite considering its color, texture and structure.

2.5 Identification of sedimentary rock

The color, mineral composition, texture and structure of many sedimentary rocks are different due to the different forming conditions, which reflect different characteristics. All of these characteristics are the main symbols for the identification of sedimentary rocks.

2.5.1 Color of sedimentary rock

The color of the sedimentary rock is the most intuitive mark of the visual identification, which is not only affected by the mineral composition of the sediment, but also affected by the palaeogeographic environment of the diagenesis.

(1) White

These kinds of rocks do not contain pigment mineral such as the rocks formed by calcite, quartz, kaolin and other minerals. There are very few pure white sedimentary rocks in nature. Most sedimentary rocks show light color or dark color.

(2) Grey and dark

The sedimentary rocks show different colors when they contain sulfide or organic matter such as carbonaceous matter and asphaltene. The higher the content of these substances is, the darker the color is.

Gray and black indicate that the rock is formed in the environment of reduction reaction or strong reduction reaction.

(3) Brown yellow, brown red, purple red and red

Most of them are the colors of continental sedimentary rocks or marine-continental transitional facies sedimentary rocks. The pigment substance of hydrated iron oxide in deposit sediment makes the sediment show brown yellow.

The pigment substance with hydrated iron oxide precipitated in the sediments makes the sediment show brownish yellow. In the later stage of diagenesis, hydrated iron oxide was converted to a high premium iron ion Fe^{3+} due to dehydration, which made the sediment show red. Red indicates that the rock is formed under the strong oxidizing environment. If it is in the condition of reduction in the later stage of diagenesis, namely Fe^{3+} is reduced to Fe^{2+}, it will appear gray green.

(4) Green

When the sedimentary rocks contain Fe^{3+} and Fe^{2+} silicate minerals such as glauconite and chlorite, they are green. This color appears due to the rock formation environment of weak oxidation or weak reduction.

2.5.2 Material composition of sedimentary rock

The composition materials of sedimentary rocks mainly include:

(1) Detrital materials

The sedimentary rock is formed by the detrital materials produced by the pre-existing rock under the action of physical weathering. Most of the detrital materials are primary mineral detritus which are chemically stable and insoluble in water such as quartz, feldspar, white mica and so on. There are some rock fragments in rocks too. Sometimes, some forms of detrital materials such as volcanic ash produced by volcano eruption can be seen in sedimentary rocks.

(2) Clay minerals

It is mainly secondary minerals formed by chemical weathering of rocks containing silica-aluminate minerals such as kaolinite, montmorillonite, hydromica and so on. The particles of these minerals are extremely fine (particle size is less than 0.005mm) and have strong hydrophilicity, plasticity and expansibility.

(3) Chemically precipitated mineral

It is the sedimentary mineral precipitated crystallization from solution due to chemical or biochemical action such as calcite, gypsum, opal and oxides or hydroxides of iron and manganese.

(4) Biological debris and organic matter

They are substances produced by the chemical changes of dead creatures or organic matter such as shells, peat and other organic matter.

Clay minerals and organic matter are unique to the sedimentary rocks and are the important characteristics of materials which differ from igneous rocks.

2.5.3 Texture of sedimentary rock

The texture of sedimentary rock is divided into four types, namely clastic texture, pelitic texture, crystalline texture and organic texture according to the characteristics of matter composition, size and shape of particle.

(1) The clastic texture is formed by cementing of the detrital material. According to particle size, it can be divided into psephitic texture, arenaceous texture and silty texture.

① Psephitic texture

The size of clastic particle is greater than 2mm. It is called brecciated texture when the debris still has edges and corners due to the short or no transport distance. It is called round psephitic texture when the debris is round or has a certain degree of roundness due to transport.

② Arenaceous texture

The size of clastic particle is between 0.05mm and 2mm. It is called coarse grained texture when the size of clastic particle is between 0.5mm and 2mm such as coarse sandstone. It is called medium grained texture when the size of clastic particle is between 0.25mm and 0.5mm such as medium sandstone. It is called fine grained texture when the size of clastic particle is between 0.05mm and 0.25mm such as fine sandstone.

③ Silty texture

The size of clastic particle is between 0.002mm and 0.05mm such as siltstone.

The texture of sedimentary rocks can also be classified according to the cementation material, which can be divided into siliceous cementation, ferruginous cementation, calcitic cementation and argillaceous cementation.

① Siliceous cementation

The rock with siliceous cementation is composed of quartz and other silicon dioxide. It has light color and high strength. The most obvious characteristic is its stiffness.

② Ferruginous cementation

The rock with ferruginous cementation is composed of iron oxides and hydroxides. It has dark color, usually appears maroon or red spots. Its strength is inferior to that of siliceous cementation. It can be weathered easily.

③ Calcitic cementation

The rock with calcitic cementation is composed of calcite and other materials such as calcium carbonate. It has light color and low strength. It can be eroded easily. It will produce bubbles when meeting diluted hydrochloric acid.

④ Argillaceous cementation

The rock with argillaceous cementation is mainly composed of fine clay minerals. It is soft and has uncertain color. It can be kneaded and scattered easily. It can often make fingers sticky dirty.

Table 2-4 shows the characteristic statistics of the clastic rocks cementation.

Cementation characteristics of clastic sedimentary rocks **Table 2-4**

Cementation	Chemical component	Main minerals	Color	Stability degree	Other characteristics
Siliceous cementation	SiO_2	Quartz Opal chalcedony Glauconite	Milk white Grey white Dark green	Perfect	Rock has high strength and hardness, and is insoluble in water
Calcitic cementation	$CaCO_3$ $Ca \cdot Mg(CO_3)_2$	Calcite Dolomite	White Grey white Light yellow Reddish color	Medium	Rock will produce bubbles with the action of dilute hydrochloric acid
Argillaceous cementation	Silicoaluminate	Kaolinite Montmorillonite Hydromica	Dark yellow Tawny	Poor	Rock is soft and easily become soften or muddy when meeting water
Ferruginous cementation	Fe_2O_3	Hematite Limonite	Red brown Tawny Brown red	Good	Rock has high strength and be weathered easily when meeting water or oxygen
Gypsum cementation	$CaSO_4 \cdot 2H_2O$	Gypsum	White Grey white	Fair	Rock has low strength and can be corroded in the condition of long term soak
Carbon cementation	C	Organic matter	Dark Dark grey	Poor	Rock has low strength and easily becomes muddy when meeting the water

(2) Pelitic texture

Pelitic texture is composed of clay mineral particles, whose size are almost less than 0.005mm. It is the main texture of mudstone, shale and other clay rocks.

(3) Crystalline texture

Crystalline texture is formed by precipitation or recrystallization of a solution. The crystalline grain produced by precipitation is very fine, which can become coarser by the action of recrystallization, but the average particle size is less than 1mm and invisible to the naked eyes. Crystalline texture is the main texture of limestone, dolomite and other chemical rocks.

(4) Organic texture

Organic texture is composed of biological body or fragment such as shell texture and coral texture, which is the main texture of biochemigenic rocks.

2.5.4 Structure of sedimentary rock

The structure characteristics of sedimentary rocks are described as follows.

(1) Bedding structure

Bedding structure is the main structure of the sedimentary rocks and one of the most important characteristics of sedimentary rocks which differ from igneous rocks.

(2) Bedding surface structure

There are often the ripple mark, rain-print, mud surface weather-shack and specimen with stylolitic structure in the bedding surface.

(3) Fossil

The fossils of plants and animals are often seen in sedimentary rocks such as the trilobite fossils and leaf fossils that are often distributed along the bedding surface. The formation geological environment and geological age of rocks can be inferred according to their fossils.

The fossils in sedimentary rocks are one of the important characteristics of sedimentary rocks Which differ from igneous rocks.

2.5.5 To name the sedimentary rock

There are many ways to name sedimentary rocks, whose details are described as follows.

(1) Named after the characteristic of its structure

The basic names of sedimentary rocks are named after their characteristic of structure, such as clastic rocks including breccias, conglomerate, coarse sandstone, medium sandstone, fine sandstone, siltstone, clay rocks including mudstone and shale, chemical and biochemical rocks including limestone and dolomite.

(2) Named after color and basic name of rock

Such as red conglomerate, dark purple sandstone, gray green siltstone, black shale, brown yellow mudstone, dark gray limestone, yellow-brown dolomite and so on.

(3) Named after the composition of mineral components and basic name of rock

Such as feldspar sandstone, quartz lithic sandstone, white mica schist, kaolinite mudstone, dolomitic limestone and so on.

(4) Named after cementation material and basic name of rock

Such as argillaceous conglomerate, ferruginous sandstone, calcitic siltstone, calcitic shale calcitic mudstone, siliceous limestone and so on.

(5) Named after thickness and basic name of rock

In the field work, the thickness of stratum should be described specifically for the sedimentary rock with actual outcrops. The thickness standard for rock formations is shown in Table 2-5.

Criteria for the thickness of sedimentary rocks **Table 2-5**

Stratum	Micro layer	Medium layer	Thin layer	Medium-thick layer	Thick layer	Huge thick layer
Thickness(cm)	<0. 2	0. 2～2	2～10	10～50	50～100	>100

The naming rules for sedimentary rocks are as follows.

(1) Clastic rock

The name format of clastic rock is" color+structure (thickness) + cementation + texture + ingredients and basic names" such as purple red medium-thick calcitic cementation fine grain quartz sandstone.

(2) Clay rock

Clay rock is often named after its structure and degree of cementation because it is difficult to identify its composition with naked eyes. Its name format is usually "color + clay mineral + foreign matter and basic name" such as brick red calcitic mudstone.

(3) Chemical and biochemical rock

Chemical and biochemical rock is often described comprehensively based on the chemical crystallization degree. Component content ratio of cementation is listed as the basis for name, and only siliceous cementation needs to be described. Its name format is often "color + structure + texture (including biological fossil) + composition and basic name" such as light grey medium-thick fine grained limestone, light yellow huge thick coarse dolomite.

2. 5. 6 Characteristics of common sedimentary rock

2. 5. 6. 1 Clastic rock

(1) Volcanoclastic rock

Volcanoclastic rock is formed due to the deposition of the debris from a volcanic eruption in the original site or other places with short transport distance on the Earth surface. It has dual natures of volcanic eruption and deposition and is the transitional type between extrusive igneous rock and sedimentary rock. Volcanoclastic rocks can be divided into three types according to the particle size as follows.

① Volcanic agglomerate

Volcanic agglomerate is mainly composed of coarse tephra, whose particle size is greater than 100mm. The cementation material is mainly volcanic ash or lava, and sometimes calcitic carbonate, silicon dioxide or clay.

② Volcanic breccias

The content of pyroclastic tephra accounts for more than 90% and its particle size is usually from 2mm to 100mm. Most of the volcanic breccias are lava breccias and often cemented by volcanic ash. It appears different colors such as dark grey, blue grey, brown grey, green and purple.

③ Tuff

Generally, tuff consists of volcanic ash and fine clastic, whose particles size is less than 2mm. It has high porosity, small gravity and easily to be weathered. Its colors mainly are grey and grey white.

(2) Sedimentary clastic rock

Sedimentary clastic rock is also called normal clastic rock. It is formed by fragmentary materials removed from pre-existing rock experiencing transport, deposition and cementation. The common sedimentary clastic rocks are described as follows.

① Conglomerate and breccia

The content of coarse clastic particles larger than 2mm is more than 50% and content of clay is less than 25% in conglomerate and breccias. They have psephitic texture. It is called breccias when its gravel is angular, whose lithology component is relatively simple. It is called conglomerate when its gravel is ground, whose lithology components are generally more complex, namely, conglomerate is often composed of a variety of rock particles and mineral particles.

② Sandstone

The particle size of sandstone is between 0.05mm and 2mm. The content of clastic particles is more than 50% and content of clay is less than 25%. It has arenaceous texture and bedded structure. It can be divided into quartz sandstone, feldspar sandstone and clastic sandstone according to the mineral composition of sand grains. It can be divided into coarse sandstone, medium sandstone and fine sandstone according to the size of sand grain. It can be divided into siliceous sandstone, ferruginous sandstone, calcitic sandstone and argillaceous sandstone according to the cementation material.

Siliceous sandstone has light color, high strength and strong anti-weathering ability. Argillaceous sandstone is generally yellow brown, has strong water imbibition and can be softened easily. The strength and stability of argillaceous sandstone are poor. The ferruginous sandstone is purple or red brown. Calcitic sandstone is white or grey white, whose strength and stability are between siliceous sandstone and argillaceous sandstone.

Sandstone is widely used as building stone in engineering because it is widely distributed and easily mined.

③ Siltstone

The content of silt particle whose size is between 0.05mm and 0.005mm is more than 50% and the content of the clay is less than 25%. It has silty texture and lamellar structure. The minerals of silt particle are mainly quartz, followed by feldspar and white mica.

The cementation is mainly calcitic cementation and ferruginous cementation. It has loose structure, low strength and poor stability.

2.5.6.2 Clay sedimentary rock

(1) Clay rock

Clay rock belongs to loose soil rock. It mainly contains clay minerals, namely kaolinite, montmorillonite and hydromica, whose total content is more than 50% and a small amount of extremely fine quartz, feldspar and so on. Clay rock is smooth and has typical pelitic texture, homogeneous material. It has strong plasticity and water absorption, which make it expand easily after absorbing water.

(2) Shale

Shale is formed due to the diagenesis of loose clay. Shale is a structural variation of clay rocks and can be splitted into thin slices along the bedding surface. Shale is named because it has lamellation.

The composition of shale is complex, which includes a small amount of quartz, sericite, chlorite, feldspar and other minerals besides all clay minerals. Shale has many colors such as gray, brown, red, light yellow, green and black.

Shale can be divided into calcitic shale, siliceous shale, ferruginous shale, carbonaceous shale and oil shale depending on the different composition mixed in the shale. All kinds of shales are easily weathered and soften with the water action. Shale has low strength.

(3) Mudstone

Mudstone has similar composition to shale, but has not lamellation structure, only massive structure.

Mudstone, which is mainly composed of kaolinite, is usually grey or yellow white with strong water absorption and easy to be soften after meeting water. The mudstone, which is mainly composed of microcrystalline kaolinite, is white, rose or light green, with a smooth surface, small plasticity, high water absorption and rapid expansion after absorbing water.

2.5.6.3 Chemical and biochemical rock

(1) Limestone

The mineral of limestone is mainly calcite, followed by a small amount of dolomite and clay minerals. It is often dark gray or light gray, but pure limestone is white. Its lithology is homogeneous.

Limestone is widely distributed and easily mined, so it is a very common building stone.

It's worth noting that limestone can produce strong bubbles prominently with the addition of cold dilute hydrochloric acid.

(2) Dolomite rock

Dolomite rock mainly contains dolomite, also contains calcite and clay minerals. It has crystalline texture. Dolomite rock can appear different colors with different impurities, but the pure dolomite rock is white.

Dolomite rock has similar engineering properties and appearance to limestone, higher strength and stability than limestone. Chemical reaction degree with hydrochloric acid can be used to differentiate the dolomite rock and limestone in the wild. Dolomite rock is one of good building materials.

(3) Marlstone

The main mineral of marlstone is calcite, but the content of clay mineral is between 25% and 50%. The mud residues appeared after reaction with hydrochloric acid. Marlstone has pelitic texture, cryptocrystalline or microcrystalline texture and massive structure. It has poor engineering geology condition for construction.

2.5.7 Identification method of sedimentary rocks by naked eyes

Bedding structure is the most important feature for sedimentary rock identification, but its sample is taken from a certain strata and so small that it cannot show the bedding structure. Therefore, the identification of sedimentary rock sample needs to observe its material composition, structure, color and so on.

(1) Clastic rocks

The content of clastic particles in clastic rock accounts for 50% of the total. Clastic particles of clastic rock can be identified by naked eyes or magnifying glass. It is important to observe the shape and size of the debris and cementation composition besides color, composition and content of debris. The conglomerate or breccia also needs to observe its cementation type.

Table 2-6 shows the observation experiences of clastic rocks.

Observation experiences of clastic rocks by particle size **Table 2-6**

Particle size *d* (mm)	>2mm	0.5~2mm	0.25~0.5mm	0.05~0.25mm	0.002~0.05mm
	Gravel grain	Sand grain			Silt grain
Name of rock	Conglomerate Breccia	Coarse sandstone	Medium sandstone	Fine sandstone	Siltstone
Observation experiences	Particle is coarse. It can be seen easily	Sand particle is obvious, which can be found by naked eyes	Particle is not obvious. You can feel tough when touch it	Its particle can be seen by magnifying glass	You can feel fine sand when you touch it by hand with water. You may dirty fingers by the mud

The main mineral of rocks with sand grain is quartz followed by feldspar and some rock debris. Quartz in sandstone is smoky gray, irregular granular, semitransparent, and has oily luster and high hardness. Feldspar in sandstone has dull color and luster, most of which is acid anorthose or orthoclase that can be weathered into kaolin easily. Its rock debris is angular or subangular.

The content percentage observation method of the quartz, felspar and rock debris in sandstone is the same as that of the igneous rocks. The subclass of the sandstone is named after the ratio of quartz, feldspar and rock debris as shown in Table 2-7.

Classification table of sandstone **Table 2-7**

Rock name	Content of mineral composition		
	Quartz	Feldspar	Rock debris
Quartz sandstone	>90%	<10%	<10%
Arkose	<75%	>25%	<25%
Lithic sandstone	<75%	<25%	>25%

The mineral composition of rock with silt grains is mainly quartz followed by white mica, feldspar and clay minerals. There is nearly not the rock debris in siltstone, which cannot be easily distinguished by naked eyes because of its small size.

(2) Argillaceous rocks

Only a small amount of extremely fine minerals or rock debris particles in argillaceous rocks can be distinguished. Almost all of argillaceous rocks are smooth and homogeneous. Argillaceous rock has strong plasticity and water absorption so that its volume can increase after absorbing water. Argillaceous rock is often dark color and soft in wet condition, light color and hard in dry condition.

It is important to observe the structure characteristics of the argillaceous rock sample.

The common argillaceous rocks are shale and mudstone. Shale can be split into very thin (its thickness is less than 1mm) foliated or paper sheet slice along the rock surface. Shale gets its name because it has microlamination that can be visible, which is named as lamellation. Mudstone only has massive texture and no obvious lamellation texture. Shale can be weathered or broken into lamellar fragments, but the mudstone can be weathered into irregular block.

Some argillaceous rocks often contain organic materials such as carbonaceous shale, oil shale and so on.

(3) Chemical and biochemical rock

Chemical rocks are formed by the chemical action of true solution, colloidal solution and biology. Its particles are completely indistinguishable. The chemical rock is dense, or its materials have certain crystalline form. Its common major minerals include calcite, dolomite, quartz, opal, and secondary minerals include glauconite, chlorite and hydrate.

Biochemigenic rock often contains paleontological relics and biodetritus.

Simple chemical experiments should be preceded to identify the chemical rocks besides observation of color, material composition, texture and structure. For example, reaction degree with dilute hydrochloric acid is often used by seeing bubbles. Table 2-8 shows the visual identification characteristics of common chemical rocks.

Visual identification characteristics of common chemical rocks **Table 2-8**

Rock	Main minerals	Color	Hardness	Reaction with dilute hydrochloric acid	Other Characteristics
Limestone	Calcite	Dark grey. Grey dark	Medium	Produce strong bubbles immediately	It is brittle. The weathering surface is often eroded. It often contains flint and opal stripes. Sometimes oolith texture and bamboo leaves texture are visible
Dolomite rock	Dolomite	Grey white. Light grey. Grey	High	Do not produce bubbles or produce bubbles very slowly	It is brittle and dense. There is often crisscrossed small channeling on weathering surface
Marlstone	Calcite Clay minerals	Light grey white. Yellow brown. Brown red. Purple	Low	Produce strong bubbles and leave muddy spots after bubble	It has thin bedding plane. It is splitted into slice easily when weathering. It often exists as a thin layer in limestone, dolomite or coal strata

The silexite is mainly composed of amorphous opal, cryptocrystalline chalcedony and fine grain quartz. The content of SiO_2 in silexite is between 70% and 90%.

Silexite is dense, hard and has conchoidal fracture. Silexite can produce spark when hammering.

Silexite often appears in the form of lentoid or concretion, and also appears in carbonate or argillaceous rocks in the form of stratiform and stripped layer.

2.6 Identification of metamorphic rock

In the long geological history, original texture, structure or mineral composition of pre-exist rock may be changed due to the action of many metamorphic factors, and then the new texture, structure or mineral composition can be formed, this kind of rock is called metamorphic rock. The factors that cause the geological characteristics change of the pre-exist rock are called metamorphic factors. The process of changing the geological characteristics under the action of metamorphic factors is called metamorphism.

2.6.1 Classification of metamorphism

The metamorphism can be divided into four types according to the main metamorphic factors as follows.

(1) Contact metamorphism

Contact metamorphism is caused by high temperature and also called thermal metamorphism.

(2) Metasomatic metamorphism

Metasomatic metamorphism is caused by the chemical active fluid and also called pneumatolytic hydrothermal metamorphism.

(3) Dynamic metamorphism

Dynamic metamorphism is caused by the dynamic pressure.

(4) Regional metamorphism

Regional metamorphism refers to the metamorphism that occurs in a larger area.

2.6.2 Mineral composition of metamorphic rock

The minerals composed of metamorphic rocks can be divided into twogroups, one group are the common minerals which can be seen in igneous rocks and sedimentary rocks such as quartz, feldspar, mica, hornblende, pyroxene, calcite, dolomite and so on. The other group is unique metamorphic minerals such as garnet, sillimanite, andalusite, topaz, kyanite, actinolite, wollastonite, chlorite, epidote, serpentine, sericite, graphite, talc and so on.

Metamorphic minerals are important markers for the identification of metamorphic rock.

2.6.3 Texture of metamorphic rock

The texture of metamorphic rock mainly includes crystalloblastic texture, pressure texture and palimpsest texture.

(1) Crystalloblastic texture

Crystalloblastic texture can be found in most metamorphic rocks, which has deeper metamorphic degree and better mineral recrystallization degree. Crystalloblastic texture is usually phanerocrystalline texture. The crystalloblastic texture can be divided into three types as follows.

① Homoeoblastic texture

All mineral grains in the metamorphic rock with homoeoblastic texture are nearly equal in size such as quartzite and marble.

② Porphyroblastic texture

The mineral grains in the metamorphic rock with porphyroblastic texture are different in size such as schist and gneiss. The mineral that forms the porphyroblast has strong crystallization ability such as garnet and tourmaline.

③ Lepidoblastic texture

Some flaky minerals are arranged parallel in a certain direction of the metamorphic rock with lepidoblastic texture such as mica schist.

(2) Pressure texture

Pressure texture can be formed due to the fragmentation and deformation of pre-exist rocks under the action of high dynamic and static pressure.

It is called cataclastic texture when the pre-exist rock is broken into blocks, and it is called mylonitic texture when the pre-exist rock is broken into fine particles under the huge pressure.

(3) Palimpsest texture

Palimpsest texture is named because the mineral composition or texture feature of the pre-exist rock is still preserved in the new formed metamorphic rock due to the uncompleted metamorphism. Palimpsest texture is a transitional texture.

For example, for argillaceous sandstone, its argillaceous cementation may become sericite and chlorite after metamorphism, but its detrital material such as quartz cannot be changed and form blastopsammitic texture.

2.6.4 Structure of metamorphic rock

The structure of metamorphic rocks is an important symbol for the identification of metamorphic rocks, which can be divided into schistosity structure and massive structure.

(1) Schistosity structure

Schistosity structure is formed due to parallel arrangement of minerals in a direction. It is the most important characteristic of metamorphic rocks which are different from igneous rocks and sedimentary rocks. Schistosity structure can be divided into four types according to their morphology.

① Tabular structure

Tabular structure is formed due to the slightest metamorphism of the metamorphic rock. When argillaceous rocks and silty rocks are under the action of crush and press to some degree, they can be stripped into thin slabs along the fracture surface that is perpendicular to the pressure direction easily; the fracture surface is called tabular structure. Some micro crystallites of mineral can only be seen on the strip surface under the microscope such as slate. Because the mineral particles are extremely fine.

② Phyllitic structure

The minerals in the metamorphic rock with phyllitic structure are basically recrystallized and arrange parallel in one direction. It is difficult to identify the mineral particles by naked eyes due to the slighter degree of metamorphism. Only the silky luster of flaky minerals and acicular minerals can be seen on the natural peel surface such as phyllite.

③ Schistose structure

Schistose structure is formed due to the deepest metamorphism and shows obvious re-crystallization of minerals. It is formed due to the directional parallel arrangement of minerals based on a needle or flake mineral. An irregular slice of the metamorphic rock with this structure can be peeled off easily along the schistosity plane such as mica schist. The schistosity usually is very thin and shows strong luster.

④ Gneissic structure

Gneissic structure is formed due to the deepest degree metamorphism and the parallel arrangement of dark and light color minerals. The metamorphic rock with gneissic structure has coarse large particles and irregular schistosity. It is hard to crack the rocks with this structure along schistosity plane such as gneiss.

(2) Massive structure

The metamorphic rock with massive structure is composed of one or several granule minerals which distribute uniformly and have undirected arrangement such as marble and quartzite.

2.6.5 To name the metamorphic rock

(1) To name the metamorphic rock with massive structure

The name mainly considers its structure and component characteristics such as coarse crystal marble, medium grained quartzite, serpentine marble and so on.

(2) To name the metamorphic rock with schistosity structure

The name format is often additional name + basic name.

The "basic name" can be represented by its type of schistosity structure, for example, the metamorphic rock with tabular structure is named after slate, and the metamorphic rock with schistose structure is named after schist.

The"additional names" can be represented by its characteristic metamorphic minerals, major mineral compositions or typical structural features.

For example, a piece of metamorphic rock with an obvious gneissic structure can be named as garnet gneiss if it contains characteristic metamorphic mineral garnet. The schist with many talc or chlorite can be named as "talc schist" and "chlorite schist" respectively.

(3) To name the metamorphic rock with palimpsest structure

When the metamorphic rock still remains the structural feature of the pre-exist rock, it is called the palimpsest structure. The name of such metamorphic rock can be named by adding "palimpsest" before its structural characteristic such as the palimpsest bedding structure, the palimpsest vesicular structure and the palimpsest amygdaloidal structure.

2.6.6 Classification of metamorphic rocks

Common metamorphic rocks can be classified according to Table 2-9 as follows.

Classification of metamorphic rocks **Table 2-9**

Rocks	Structure	Typical rocks	Main minerals	Pre-exist rocks
Schistosity rock	Tabular structure	Killas	Clay minerals, mica, chlorite, quartz, feldspar, etc	Clay rock, clay siltstone and tuff
	Phyllitic structure	Phyllite	Sericite, quartz, chlorite, etc	Clay rock, clay siltstone and tuff
	Schistose structure	Schist	Mica and quartz are the main minerals followed by amphibole in mica schist. Talc and sericite are the main minerals followed by chlorite and calcite in talc-schist. Chlorite and quartz are the main minerals followed by talc and calcite in chlorite schist	Clay rock, sandstone, intermediate acid igneous rock. Ultrabasic rocks and dolomitic marlstone. Intermediate-basic igneous rock and dolomitic marlstone
	Gneissic structure	Gneiss	Feldspar, quartz and mica are the main minerals followed by hornblende and sometimes garnet in granitic gneiss. Feldspar, quartz and hornblende are the main minerals followed by mica and sometimes garnet in amphibole gneiss	Intermediate acid igneous rock, clay rock, siltstone and sandstone
Massive rock	Massive structure	Marble	Calcite and dolomite	Limestone and dolomite
		Quartzite	Quartz, sericite and white mica	Sandstone, siliceous rock
		Serpentinite	Serpentine, talc, chlorite and calcite	Ultrabasic rocks

2.6.7 Characteristics of common metamorphic rocks

2.6.7.1 Schistosity rock

(1) Gneiss

Gneiss has gneissic structure and holocrystalline phanerocrystalline crystalloblastic texture or palimpsest texture. Its crystalline grain size is so big due to mineral recrystallization that it can be identified by naked eyes.

The light colored minerals are mostlygranular, and mainly contain quartz and feldspar. The total content of these two minerals is greater than 50%, and the content of feldspar is greater than 25%.

The dark minerals are mostly acicular or flake. They mainly contain hornblende, black mica and pyroxene, whose total content is greater than 30%. Sometimes they contain a small amount of metamorphic minerals such as garnet.

The further name for gneiss depends on the main mineral components such as granitic gneiss, diorite gneiss and so on.

(2) Schist

Schist has schistose structure and crystalloblastic texture. It is mainly composed of some flaky minerals such as mica, chlorite, talc, quartz, hornblende, and some metamorphic minerals such as garnet.

It's worth noting that schist contains no or only trace amounts of feldspar.

Sometimes there is no clear demarcation between gneiss and schist, because there is a gradual transition between them. But most gneiss contains a considerable amount of feldspar, schist does not. So it is customary to differentiate gneiss from schist according to the content of coarse feldspar.

Schist is very similar to phyllite, but its metamorphism degree is deeper than that of phyllite.

Schist may be named as mica schist, chlorite schist, amphibole schist, talc schist, garnet schist and graphite schist according to the different types of metamorphic minerals in schist.

(3) Phyllite

Phyllite has phyllitic structure and palimpsest texture or microscopically flaky crystalloblastic texture. It is often gray, green, brown or black. Its main minerals are fine grain or flaky quartz, sericite, chlorite and so on, which arrange in one direction along the schistosity surface to form unique phyllitic structure. There is strong silky luster on the schistosity surface and wrinkles on the section plane perpendicular to schistosity surface.

The metamorphism degree of phyllite is deeper than killas.

(4) Killas

Killas has tabular structure and palimpsest muddy texture or dense cryptocrystalline texture. It is often dark gray, black, sometimes gray green, purple or red. Its main minerals are clay minerals, which are difficult to be identified by the naked eye, and occasionally a small amount of sericite or chlorite ramentum.

Killas is dense, uniform and hard. It has flat tabular structure and shows dull luster. Killas can be split into flat slabstone along the tabular structure surface. Killas can be widely used as building stone such as pavement engineering.

2.6.7.2 Massive rock

(1) Marble

Marble is formed by the recrystallization of limestone or dolomite under the action of contact metamorphism or regional metamorphism. Marble has massive structure and homeocrystalline crystalloblastic texture.

The pure marble is white. It may be grey white, light red, light green or even dark when containing impurities. It has so small hardness that knife can easily leave a scratch on it.

Marble that is mainly composed of calcite produces bubbles strongly when meeting the cold dilute hydrochloric acid. The powder of marble that is mainly composed of dolomite produces bubbles weakly when meeting the cold dilute hydrochloric acid.

(2) Quartzite

Quartzite is formed by the regional metamorphism or contact metamorphism of granite. The pure quartzite is white. It may be grey, light yellow, red brown or purple when containing impurities. Quartzite has homeocrystalline crystalloblastic texture and dense massive structure. It generally shows strong oily luster. It mainly contains quartz and occasionally a small amount of feldspar, mica, chlorite, hornblende and pyroxene.

Quartzite is hard and brittle.

Sometimes it's difficult to differentiate quartzite from marble, the difference between which is that the marble can produce bubbles with hydrochloric acid and has less hardness than quartz.

(3) Mylonite

Mylonite has obvious mylonitic texture and massive structure. Composition materials arrange in one direction to form band or different color stripe, between which there is often rigid materials with unequal size augen shape, lenticular shape or lentoid shape such as quartz, feldspar or their aggregate. Sometimes, mylonite contains a small amount of sericite, chlorite, serpentine, epidote or other minerals. Mylonite can only be found in larger scale faults.

2.6.8 Identification method of metamorphic rocks

When identifying metamorphic rocks, the structure of metamorphic rocks should be analyzed first, and then the main mineral composition and characteristic minerals should be observed. Finally, the rock can be named.

2.6.8.1 Structure of metamorphic rock

The structure of metamorphic rock has special significance, which is not possessed by igneous rock and sedimentary rock. The structure of metamorphic rock should be analyzed in order to distinguish the schistosity structure such as killas, phyllite, schist and gneiss, and massive structure such as marble, quartzite, mylonite and so on.

(1) Tabular structure

The soft rock such as shale will appear a group fracture surfaces that are parallel to each other, namely cleavage surfaces when it is under the action of regional low-temperature dynamic metamorphism.

The cleavage surface is always flat and smooth, sometimes contains a small amount of sericite, chlorite and so on.

The rock with tabular structure is usually formed due to the lower degree regional metamorphism such as killas. It only has a small amount of newly formed minerals.

(2) Phyllitic structure

The rock with phyllitic structure has finer mineral particles that are invisible to naked eyes due to the low degree metamorphic recrystallization action. Only the cleavage surface shows a strong silky luster, which is caused by the dense arrangement of sericite and chlorite. Usually, there are many small wrinkles on the cleavage surface. The micro cleavage is often small and thin. The new metamorphic mineral particles arrange continuously or form micro wrinkles under a microscope.

(3) Schistose structure

The rock with schistose structure is mainly composed of flake minerals such as mica and some granular minerals, whose schistosity surface may be flat or wave bending.

It is termed as lineation when there are only granular minerals arranged in one direction in the rock. It is termed as fold structure when the schistosity crumples up during the metamorphosis. In some metamorphic regions, the rock may appear secondary schistosity due to more than one time tectonism. There is angle difference between the secondary schistosity and the early schistosity, which reflects the different directions of the two tectonic stresses.

(4) Gneissic structure

Gneissic structure is called gneissic schistosity. In addition to the granular minerals in the rock, there are a certain number of flaky minerals arranged in one direction and columnar minerals arranged discontinuously.

If the distribution of flaky and columnar minerals is more concentrated and continuous, it can be a banded structure with different particle size or different color. For example, it may appear striated structure when the dark colored columnar minerals distribute discontinuously in light colored granular minerals.

(5) Massive structure

The mineral composition and structure in the rockwith massive structure are very uniform and dense such as quartzite, marble and so on, which do not show directional arrangement and are dense.

(6) Palimpsest structure

It is termed as palimpsest structures when the metamorphic rocks still retain the structural characteristics of the pre-exist rocks.

Palimpsest bedding structure is the common structure in the metamorphic rocks formed by sedimentary rocks. In addition, there are also palimpsest crossing bedding structure, palimpsest ripple mark, mud cracks and dead creatures.

The common structures in metamorphic rocks formed by igneous rocks are palimpsest vesicular structure, palimpsest amygdaloidal structure, palimpsest pillow structure, strip structure, etc.

2.6.8.2 Texture of metamorphic rock

According to the crystallization habits and morphology of the minerals in metamorphic rocks, granular crystalloblastic texture is often found in marble and quartzite, flaky crystalloblastic texture is usually found in crystalline schist and gneiss, fibrous crystalloblastic texture is mostly found in schist.

2.6.8.3 Major minerals and characteristic minerals of metamorphic rock

(1) To distinguish light colored and dark colored minerals

If the rock is dominated by light colored minerals, there is more quartz and less feldspar or no feldspar such as quartz schist, mica schist and so on.

If the rock is dominated by dark colored minerals and has a large amount of feldspar, it is usually gneiss such as amphibole plagiogneiss, mica gneiss and so on.

(2) To identify the characteristic minerals

Chlorite and sericite often appear in the slight degree metamorphic belt, the kyanite represents the medium degree metamorphic belt and the sillimanite represents the deepest degree metamorphic belt. All of these minerals are called the standard marker mineral.

2.7 The basic identification method of three kinds of rocks in the field

Three kinds of rocks are not isolated in nature, but depend on each other and can transform under certain condition. There are great differences of material composition, texture and structure between pre-exist rocks and the newly formed rocks. The inherent characteristics of three kinds of rocks are summarized as shown in Table 2-10.

Characteristics of three kinds of rocks **Table 2-10**

Characteristic	Rock group		
	Igneous rocks	Sedimentary rocks	Metamorphic rocks
Mineral composition	It is composed by primary mineral, which is complex such as quartz, feldspar, hornblende, pyroxene, olivine, black mica and so on	It is composed by the primary minerals including quartz, feldspar, white mica and secondary minerals including calcite, dolomite, kaolinite, glauconite and so on	It is composed by the minerals of pre-exist rocks and metamorphic minerals including sericite, garnet, chlorite and so on
Texture	Granular crystalline texture. Porphyritic texture. Colloidal texture	Clastic texture. Pelitic texture. Biochemical texture	Crystalloblastic texture. Palimpsest texture. Pressure texture
Structure	Rhyotaxitic structure. Vesicular structure. Amygdaloidal structure. Massive structure	Bedding structure. Biological fossils	Schistosity structure. Gneissic structure
Occurrence	Many of them appear in intrusive bodies, and few are eruptive rocks, which are irregular	It has regular layers	It is determined by the occurrence of pre-exist rocks
Distribution	Granite and basalt are the most widely distributed	Clay rocks are the most widely distributed followed by sandstone and limestone	Regional metamorphic rocks are the most widely distributed followed by contact metamorphic rocks and dynamic metamorphic rocks

Identification of a rock in nature can be carried out according to the following steps.

(1) The first step is to identify the type of the rock according to its general appearance characteristics and structure.

The structure of the rock can reflect its genetic type according to its appearance. For example, the rock with vesicular structure, amygdaloidal structure or rhyolitic structure usually belongs to the extrusive rock in igneous rock, the rock with bedding structure or bedding plane structure usually belongs to sedimentary rock, the rock with tabular structure, phyllitic structure, schistose structure or gneissic structure usually belongs to the metamorphic rock.

It is worth noting that all of the three kinds of rocks have "massive structure" such as the quartz porphyry sample in igneous rock, quartz sandstone sample in sedimentary rock, quartzite sample in metamorphic rock. It is difficult to distinguish them according to their appearances, so the rocks need to be identified combining with its structure characteristics and mineral composition at the same time.

Quartz porphyry has porphyaceous texture of igneous rock, whose phenocryst and ground mass minerals are connected by crystallization action. Quartz phenocryst in quartz porphyry has a certain crystal appearance, which is columnar or granular.

Quartz sandstone has the clastic texture of sedimentary rock, whose debris is connected by cementation with each other. The particles in the quartz sandstone are uniform in size and rounded, whose glassy luster has disappeared. Little concave pit appears in the place where cement is not fastened after the separation of quartz granule and cementation due to hammering or cutting.

Quartzite is formed due to the recrystallization metamorphism. It is usually dense, hard and brittle. It has quartz granule that cannot be identified by naked eyes.

The sedimentary rock is often exposed to steep and slight slopes alternately. A slight slope is usually formed along the bedding structure surface.

In the geomorphology, if there is no influence of tectonic movement, igneous rocks often form undulating terrain. There is no steep and slight slopes arranged alternately in igneous rocks area like sedimentary rocks area.

(2) The second step is to classify the rock further by observing its structure carefully.

For example, most plutonic intrusive rocks of igneous rocks have holocrystalline texture, phanerocrystalline texture and equigranular texture, but the hypabyssal intrusive rocks often have a porphyritic crystalline texture.

Sedimentary rocks include the clastic rocks such as conglomerate and sandstone, clay rocks such as shale and mudstone, and biochemical rocks such as limestone.

(3) The third step is to observe the material composition of the rock including minerals, detrital material and cementation material with magnifying glass and knife.

The mineral composition and chemical composition of rocks are indispensable to the classification and naming of rocks, especially for igneous rocks. For example, both porphyry and porphyrite belong to hypabyssal intrusive rock, whose main difference is the mineral composition. The phenocryst minerals of porphyry are mainly orthoclase and quartz, but the phenocryst minerals of porphyrite are mainly anorthose and dark colored minerals such as hornblende and pyroxene.

The secondary minerals in sedimentary rocks such as calcite, dolomite, kaolinite gypsum and limonite are not likely to exist in fresh igneous rocks.

Sericite, chlorite, talc, asbestos, garnet are unique to metamorphic rocks. Therefore the classification of rocks can be preliminarily determined according to the composition analysis of some metamorphic minerals.

(4) The fourth step is to name the rocks.

If the rock is composed by multi minerals, the rock will be named as "the name of mineral with most content + basic name of the rock", and the name of other secondary minerals will be arranged in turn to the left according to their content. For example, hornblende plagiogneiss means that it mainly contains anorthose and contains a considerable amount of hornblende at the same time.

The name meaning of igneous rocks and metamorphic rocks is similar.

For example, the rock can be identified as sedimentary rock when it has obvious layers in appearance and developed sedimentary bedding surface in the wild. From further observation we can know the rock belongs to clastic sedimentary rock because it is composed of detrital material and cementation material. At last, the rock can be named after coarse-grained feldspar quartz sandstone because it has coarse grained texture and its clastic material is mainly quartz and feldspar.

It is important to note that it is difficult to identify many minerals in the rock sample by naked eyes such as igneous rock with cryptocrystalline texture or vitreous texture, sedimentary rock with pelitic texture or chemical texture, and some metamorphic rocks that are composed of crystalline fine material or noncrystalline material. In general, it only can be judged according to the color, hardness, specific gravity and reaction degree with hydrochloric acid.

Igneous rocks with most dark color minerals often belong to basic rocks and with most light color minerals often belong to acid rocks.

Most of the more solid sedimentary rocks are siliceous cementation rocks. The sedimentary rock with greater specific gravity is mostly the rock containing iron and manganese. If the rock can produce the reaction when meeting hydrochloric acid, it must be carbonate rock.

Biological fossil is the typical characteristics of sedimentary rock, but there is no fossil in igneous rocks and metamorphic rocks.

2.8 Internship report

Please observe the characteristics of rock samples and fill the observation results in Table 2-11.

Table 2-11

Identification table of unknown rocks samples

Number of samples	Color	Mineral composition	Texture	Structure	Identification characteristic	Name of rock

2.9 Thinking questions

Please compare the similarities and differences of the following groups of minerals or rocks.

① Fibrous gypsum and asbestos.

② Calcite and dolomite.

③ Clay rock and marlstone.

④ Killas and shale.

⑤ Siliceous limestone and dolomite.

Chapter 3 Use of geological compass instrument

3. 1 Structure of geological compass instrument

The spatial positions of strata, folds, faults and broken planes are characterized by their attitude including strike direction angle, dip direction angle and inclination angle. The attitude can be determined by geologic compass or by means of drawing or calculating. The geological compass is an essential portable instrument in the field geology work, and it can be used to determine the three attitude elements of any observation surface (strata layer, fold axis, fault plane, joints, etc.) and the various tectonic elements of igneous rock and the occurrence of orebody.

This chapter requires understanding the structure of geological compass instrument, understanding its main uses and mastering the method of measuring the occurrence of strata (faults) under different conditions by using a geological compass instrument.

Geological compass is a lot of design but the structure is basically the same, commonly used circular-basin geological compass by the magnetic needle, dial, inclinometer, leveling and other parts of the composition, as shown in Figure 3-1.

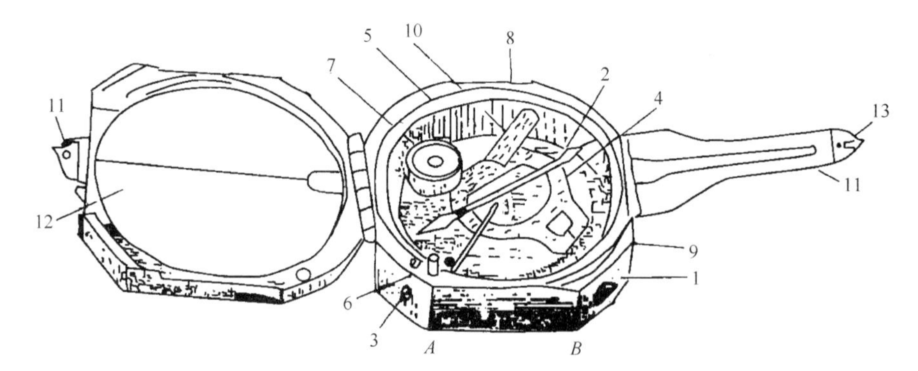

Figure 3-1 Geological compass instrument

1—chassis; 2—magnetic needle; 3—disc correction screw; 4—inclination meter; 5—disc; 6—magnetic needle brake and inclination instrument brake; 7—leveling bubble; 8—azimuth dial; 9—inclination angle dial; 10—inclination meter leveling bubble; 11—folding sighting device; 12—glass mirror; 13—observation hole

3. 1. 1 Compass

Compass can be used to determine the direction such as the direction of the target, the strike and dip direction of the strata. Its main components are as follows.

① Sighting device: There are two sighting devices in compass, namely, the front one and the rear one. The front folding-type sighting device is also referred to as the north standard. The sighting device is made up of sight bead and observation holes, used to target objects.

② Azimuth dial (horizontal dial): The counterclockwise direction is engraved with 360° azimuth value, a few compass have quadrant angle, dial is divided into four quadrants, each quadrant is 0° ~ 90°, of which the South and north are 0°, east and west are 90°.

③ Magnetic needle: It is generally the diamond-shaped steel needle with middle width and both sharp sides. It mounted in the center of the chassis on the thimble, can freely rotate. In the northern hemisphere, the compass used in the country, wrapped around the end of the copper wire is the magnetic south needle, the other end is the magnetic north needle. When the magnetic needle is static horizontally, the North needle points to the corresponding degree on the dial, which is the magnetic azimuth pointed by the North Mark, and if the declination correction on the compass, it is the true azimuth angle.

Geological compass should be used to tighten the brake screws, the magnetic needle lift pressure on the cover glass to avoid the needle cap and the tip of the collision, so as to protect the top tip, prolong the use of the compass time. In the measurement of the relaxation of the brake screws, so that the magnetic needle free swing, the last static when the needle is the direction of the needle meridian.

④ Magnetic needle brake: It is used to brake the magnetic needle, so as to be read easily. When the compass is not used, the needle and the top of the support shaft separate, so that the rotation of the friction site can be protected.

⑤ Azimuth dial correction screw: The azimuth dial can be rotated by a slight rotation to achieve the purpose of correcting declination.

⑥ Leveling bubble: Used to help compass position level, when the bubble is centered, the compass chassis is in a horizontal position, the magnetic needle flexible rotation, to its static after the direction of the azimuth or quadrant to indicate the degree of angle.

3.1.2 Tilt meter

The tilt gauge is used to measure inclination angles such as strata inclination. The main components of the tilt meter include tilt dial, tilt indicator, tilt meter level bubble and tilt instrument brake (located on the back of the chassis), where in the tilt indicator, tilt meter level bubble and tilt instrument brake are fastened together. The compass instrument chassis long edge to the measured tilt line, when the pull brake to the tilt meter leveling bubble center, the pointer refers to the scale of the dial is the angle of the tilt line.

Compass Azimuth dial in the counterclockwise direction of 360 °, which is the direction of azimuth and the exact opposite. This is because in the specific direction of measurement, the operation of the rotation of the compass, and the magnetic needle is always pointing to the south, magnetic north pole, relative to the dial is exactly the opposite direction, so counterclockwise to be in the dial on the measurement of the direction of the (magnetic) azimuth value.

3.2 Use method of geological compass instrument

3.2.1 Declination correction

Declination correction is required before geologic compass is used, because the north and south poles of the geomagnetic field are not exactly the same as the north and south polar positions, that is, the magnetic meridian does not coincide with the geographical meridian, the magnetic north direction of the Earth is inconsistent with the direction of the point, the angle between the two directions is called declination.

The north side of the magnetic needle at some point on the earth is called the East Side, and the west side is the opposite.

Declination all over the earth is scheduled to be calculated and published for reference. If the declination of a point is known, the relationship between the magnetic azimuth α and the North azimuth α of a line is $A=\alpha' \pm$ declination. The application of this principle can be used to correct the declination and can rotate the compass tick spiral, so that the horizontal dial to the left or right (declination east to right, west to the left), and the angle of the compass chassis north-south tick marks and horizontal dial line equals declination. After the correction the measurement of the reading is true azimuth angle.

3.2.2 Azimuth measurement steps

Azimuth measurement is to determine the relative position relationship between the object and the tester, i.e. the azimuth angle of the object (azimuth is the angle from the meridian clockwise direction to the line).

When measuring, loosen the brake screw so that the plate of the object is pointed to the measuring object. Even if the compass to the north end of the object, the southern end of their own, to aim at, so that the object, the plate small hole, cover glass and the mesh of plate holes and so on line, while making the chassis leveling bubble center, and when the needle is still at rest refering to the north needle the degree of the target is measured in the direction of the object (if the pointer is not static, can be read one-second of the minimum degree of the magnetic needle swing, the measurement of other elements reading is the same).

If the measured plate is aimed at the measuring person (the southern end of the compass is aiming at the object), the North needle reading indicates that the tester is in the relative direction of the object, at which point the reading of the compass is the relative direction of the object in the test, compared with the former because the compass aims at the object two times, the north and south ends are reversed. Therefore, the relative position of the measured object and the measured person is affected.

In order to use the reading point of the north needle and reading the compass produced by the confusion, the plate should be pointed to the direction of constant reading of the north needle; the reading is to be measured by the azimuth angle.

3.2.3 Compass use precautions

(1) Magnetic needle and thimble, agate bearing is the most important part of the instrument, should be used carefully, keep clean, lest affect the sensitivity of the magnetic needle. When not in use, the instrument should be shut down, after the instrument is closed, through the switch and the action of the lever will automatically lift the needle, so that the thimble and agate bearings out, so as not to grind bad thimble.

(2) Do not easily remove all hinges, so as not to loosen the impact of precision.

(3) Hinge rotation part should often order some watch oil to avoid dry grinding and broken.

(4) The instrument should try to avoid high temperature exposure, so as to avoid leakage of blisters failure.

(5) When the compass instrument is not used for a long time, it should be put in the ventilated and dry place, lest mildew.

3.3 Strata attitude measurement

The spatial position of rock strata depends on its formation factors, and the strata attitude factors include the strike direction, dip direction and inclination angle of strata. The measurement of strata attitude is one of the most basic working methods in field geology work, and must be mastered skillfully.

The geological compass is used to measure strike direction, dip direction and inclination angle of strata at the rock level, as shown in Figure 3-2. When the geological compass is used in a certain area, the compass instrument declination should be corrected according to the declination of the area first. The western part of China is declination, so the declination degree should be added to the east, and in the middle part of China the declination degree should be subtracted from the declination. After declination correction, the azimuth angle measured by geological compass in this area is true azimuth.

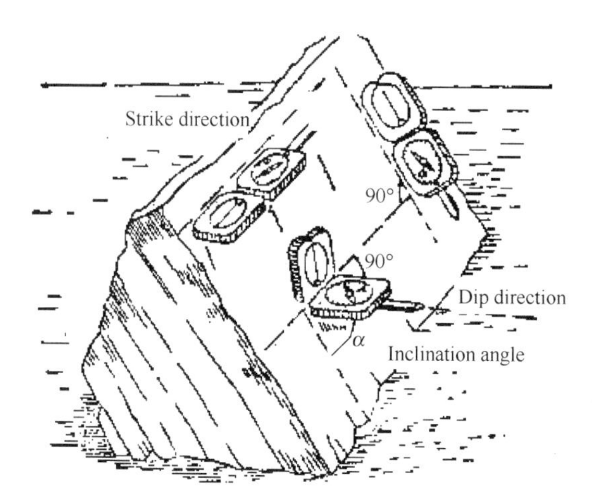

Figure 3-2 the measuring method of strata formation attitude

3.3.1 Determination of the strike direction of strata

The strike direction of strata is the intersection lines direction of strata with horizontal plane at any heigh.

When measuring the strike direction of rock strata, a long side and a plane of the compass should be pressed tightly, then the compass will be rotated to center the blister of the chassis level, and the calibration value of the magnetic needle on the dial shall be the azimuth value of the two orientations of the strata. Such as NE30° and SW210° can represent the direction of the rock strata. Among them, the corresponding scale value of the magnetic north needle is the direction of the side of the folding sight, and the corresponding scale value of the magnetic south needle is opposite another direction.

3.3.2 Determination of the dip direction of strata

The dip direction of strata refers to the projection of the downward maximum inclined direction line of strata to the horizontal plane, which is always perpendicular to the strata.

When measuring the dip direction of strata, a short edge of the compass is pressed against the rock layer, and the north end of the compass or the things plate is directed toward the rock strata. The southern end of the compass rotates, so that the chassis level bubble is center, and when the needle is still, the magnetic north needle refers to the scale that is the orientation of the rock strata.

If it is difficult to measure on the top of the rock, it can also be measured on the bottom surface of the rock, still use the plate to point to the direction of the rock strata, the compass north end close to the bottom, read the north needle can be, if the measurement of the bottom of the north needle reading refers to obstacles, then use the southern tip of the compass close to the bottom of the rock, reading the compass can also be.

Between the strike direction and dip direction we only need to measure one. If only the strike direction is measured, the dip direction of strata can be converted according to the actual inclination of the strata. The conversion method is: After reading out a strike direction angle of the strata, if the direction of 90° is rotated clockwise to dip direction, then the azimuth value of this direction is inclined to add 90°; if this trend counterclockwise rotates 90° to dip direction, then the azimuth value reduced by 90°.

3.3.3 Determination of dip angle of strata

The dip angle of rock strata is the maximum angle between the strata plane and the imaginary horizontal plane, i. e. the true dip direction, which is measured by the true dip direction along the strata, and the dip angle measured in other directions. The apparent dip angle is less than the true dip angle, that is, the angle between the true inclination line and the horizontal plane at the stratum level is true inclination, and the angle between the view inclined line and the horizontal plane is the apparent inclination angle. The true inclination direction of the field resolution is very important, and it is always perpendicular to the strike direction. In addition, you can let pebbles in the level of rolling or dripping at the level of flow, the rolling or flow direction is the true inclination direction of the plane.

When measuring the strike direction or dip direction, the compass is drawn to the contact line at the rock level. The real inclined line can be drawn on the rock level by using the vertical relation of the long and short edges of compass.

When surveying strata in the field, it needs to be measured at outcrop, not on rolling stones, so that the outcrops and rolling stones are distinguished. The difference between outcrop and rolling stone is mostly observation and recourse, and good at judgment.

When measuring the attitude of rock layers, if the strata are uneven, the notebook can be flat on the rock layer for measurement.

3.3.4 Attitude measurement at steep slope

It is more convenient and accurate to measure the attitude of the shape at steep rock level as following steps.

(1) The compass mirror cover is close to the rock layer and adjusts the bubble to the level, and the scale value of the magnetic north needle is the value of the inclination azimuth angle.

(2) The long edge of the mirror cover is the true inclination line.

(3) The inclination angle can be measured if the compass side is made to coincide with the long side and the true inclined line.

3.3.5 Attitude representation method

The attitude of strata is usually recorded in the azimuth method, usually only the dip direction angle and dip angle. If a rock layer is measured to be 310° in strike direction, 220° in dip direction and has 30° dip angle, it is recorded as SW220°∠30°, or 220°∠30°, or N50°W∠30°SW, or S50°E∠30°SW.

3.4 Internship report

Please measure the strata to fill in Table 3-1.

Strata attitude statistics **Table 3-1**

Location and name of the strata	Strike direction	Dip direction	Dip angle

3.5 Thinking questions

How to measure the attitude of strata, faults and folds? What are the similarities and differences?

Chapter 4 Field observation and analysis of geological structure

Geological structure is the production of crustal movement, because of the great stress in the Earth's crust, the upper strata of the Earth's crust are deformed and modified under the long-term effect of the ground stress, which forms the traces of tectonic change such as the folds and faults that are often encountered in the field. All kinds of permanent deformations and displacements left over by tectonic changes in rock mass are called geological structures.

The sizes of the geological structures are different. The large tectonic belt is thousands of kilometers long such as the San Andreas Fault in the United States. The small geological structure includes the formation of magmatic rock and so on. In the same region, there are often different scale and different types of tectonic system, so the regional geological conditions are complicated, but large and complex geological structure is shown always by some small basic tectonic form in a certain way.

The different structures in the field are observed from the small structures visible on the outcrop, by observing and describing its morphology, structural elements and types of occurrence, the spatial combinatorial relationship and the developmental order of time were identified, then the geological phenomena of different outcrops were connected and the relationship between them and large structures was determined.

4.1 Field observation of uniclinal structure

As the crustal movement causes tectonic changes in the original horizontal strata to form a tilted rock stratum, this is the simplest tectonic change and the most common state of layered rock. When the angle between the tilted strata and the earth level is between $10° \sim 70°$, it is called a uniclinal structure, and the landform composed of uniclinal structures is called Cuesta, and the mountain ridge formed by inclined strata greater than $40°$ is called Hogback.

Uniclinal structure may be a part of a geological structure such as a wing of a fold or a wall of a fault, or a regional tilt caused by an uneven settlement in the crust.

In the area of uniclinal structure strata, special attention should be paid to the relationship between the strike direction of the route and the attitude of the strata. As the strike direction of the route moves in line with the strike direction of the strata, the footslope of the reverse slope often has loose slope wash or colluvium, which is unfavorable to the stability of road subgrade. If the angle of the uniclinal structure in dip slope is greater than 45°, and the thin layer, or sandwiched with weak strata, it is easy to appear slope collapse or landslide. When the strike direction of the route is perpendicular to the strike direction of the rock strata, if there is no inclination to the joints of the roadbed, a stable high steep slope can be formed, and the slope stability is between the two cases mentioned above when the route direction is inclined to the rock stratum.

4.2 Field observation of fold structure

Symmetry repetition of strata is the basic method to determine folds. In most cases, the sign layer should be selected and identified in a certain area, and it should be traced to determine whether there is a turning point on the section and whether there is a prone or uplifted plane. The primary sedimentary structure of sedimentary rock should be studied in order to determine whether it is normal level or inverted layer in the area where sedimentary rock develops or where tectonic deformation is strong.

A bend in a fold structure is called a fold, and the basic form of the fold consists of-syncline and anticline. The syncline is the downward bend of the strata, and the anticline is a curved upward uplift of the strata. It should be noted that the field observation of the fold structure cannot be simply based on terrain fluctuations, and attentions have to be paid to the inverse tectonic landform such as Anticline Valley and Syncline Mountain. There are two kinds of traverse method and recourse method in the field observation, which are mainly through the traverse method, supplemented by recourse method and interspersed with the use.

(1) Traverse method: It is easy to understand the occurrence, sequence and new and old order of strata by observing perpendicular to the strike direction of strata along the selected survey route. If the strata in the route are repeated regularly, it will be the fold structure, and then, according to the sequence and its new and old orders, the judge is the anticline or the syncline; then the attitudes of the both wings and the relationship between the both wings and the axial plane should be analyzed, and the forming type of the folds is judged.

(2) Recourse method: Observation can be preceded parallel to the strike direction of the strata along the selected survy route. This method is convenient to find out the extension direction and the structure change of the fold. Generally, the fold with both wings parallel on the plane is the horizontal fold; the fold with both wings closed at the turning ends is the plunging fold.

4.3 Field observation of fault

Fault is the fracture structure with obvious relative displacements along the fault surface on both sides of the faults. The fault is widely developed, the scale is very different, the big fault extends hundreds of kilometers or even thousands of kilometers, and the small faults can be seen on the hand specimen. Some faults cut through the lithosphere of the Earth's crust, while others develop in shallow layers of the surface. Fault damages the integrity of rock strata, not only has a great influence on the permeability, stability and regional stability, but also is a good channel and gathering place for groundwater movement, and is often endowed with abundant groundwater resources in the vicinity of large faults or fault areas.

4.3.1 Field identification of fault

If in the field we can directly see the fault level, the fault can be confirmed directly. However, the fault surface is often not directly observed, some other signs need to be found to identify the fault.

4.3.1.1 Geomorphic Signs

(1) The place where the fault passes is often expressed as a depression or a river valley, which is the result of the broken rock easily eroded and cut by the fault, but it cannot be considered as "every ditch must be fault".

(2) The ridge is broken and staggered, valley overfall waterfall, valley direction abrupt transition, etc.; it is likely fault dislocation in the topography of the reflection.

(3) The new faults in the Times often form cliff steep (fault cliff), and the fault cliff is weathered and eroded to form the surface topography of the fault, as shown in Figure 4-1.

Figure 4-1 Schematic diagram of the fault triangle surface

a—Fault cliff denudation into a gully;

b—Gully expands to form a triangular surface;

c—Continues to erode and the triangular surface disappears

4.3.1.2 Structure sign

In the formation process of faults, the associated structures are formed by squeezing and dislocation of the two disk blocks, such as rock bend, fault breccia, mylonite, fault gouge and slickenside, as shown in Figure 4-2.

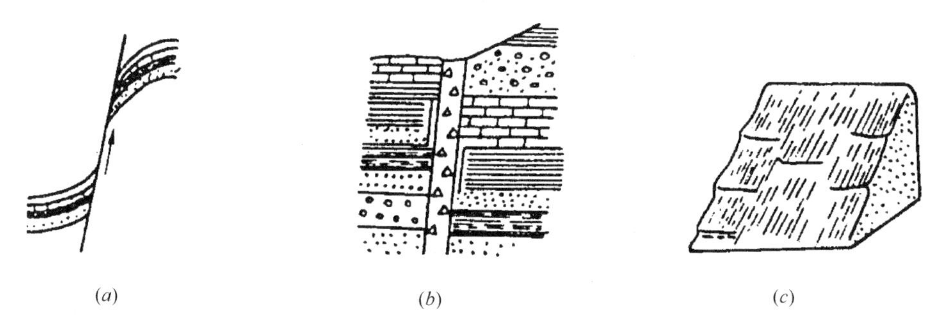

(*a*) (*b*) (*c*)

Figure 4-2 Structural sign of fault

(*a*) rock bend; (*b*) fault breccia; (*c*) fault gouge

The rock bend of strata is the bending of the strata on both sides of the fault plane due to the relative dislocation, which is formed in the soft rock such as shale and schist. When the two walls of fault are strongly squeezed relative dislocation, the fine mud which is milled along the fault plane is called the fault gouge, and if it is broken into different size, the angular gravel is called the fault breccia. When two walls are relatively dislocation, the grooves in the fault plane are parallel to each other, which are called the slickenside.

4.3.1.3 Stratigraphic Markers

Stratigraphic markers are reliable evidences for determining the existence of faults, such as stratigraphic repetition, stratum loss, dikes or veins disconnected, etc., as shown in Figure 4-3.

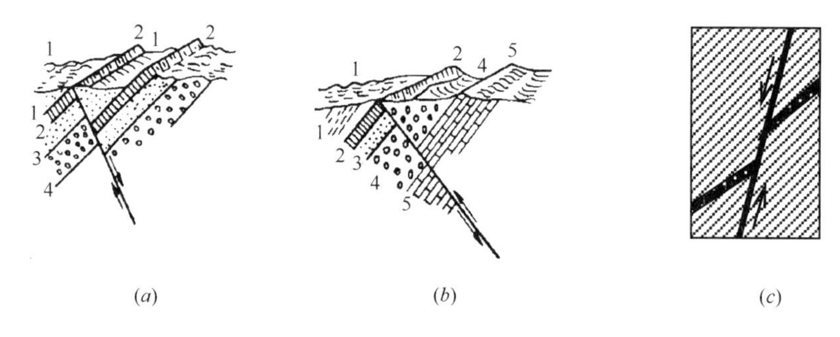

(*a*) (*b*) (*c*)

Figure 4-3 Fault Signs

(*a*) stratigraphic repetition; (*b*) stratum loss; (*c*) dike disconnected

4.3.1.4 Other Signs

If the spring or hot springs are linear out, there may be faults; when the fold structure is cut across the fault, the outcrop width of the core strata on both sides of the fault plane is different, that is, the abrupt change of the stratum width of the fold core is also the sign of distinguishing the fault.

4.3.2 Field characteristics of fault

4.3.2.1 Normal Fault

The upper wall of the normal fault faces the downward movement along the fault, while the footwall slides relatively upwards. The attitude of normal fault is generally steeper, its dip angle is more than 45°, and 60°~70° is the most common, the fault plane of large normal fault that faces the deep underground often become gentle. In the normal fault zone, the rock fragmentation is relatively less strong, the breccia is more angular, and the complex small folds are usually not formed by the intense squeezing.

4.3.2.2 Reverse fault

The upper wall of the reverse fault slides along the fault plane, and the footwall slides relative downward. The attitude of the reverse fault is generally gentle, and its dip angle is usually around 30° or more gentlely. The reverse fault often shows a strong crushing phenomenon, forming breccia, clastic rock and super fractured rock. Along the reverse fault also often appear cleavage, joint, shear zone or a variety of complex ruffles, these structures are often banded arrangement, or interwoven interspersed. Rock formations on both walls of the reverse fault often have strong deformation characteristics.

4.3.2.3 Parallel displacement fault

The two walls of the parallel displacement fault move relative to each other along the strike direction of the fault, because the fault plane is usually steep upright, the effect of the normal fault or the reverse fault is consistent with that in the transverse section, and sometimes it has a certain degree of sliding along dip direction, so some large parallel displacement faults are often mistaken for normal faults or reverse faults. Large-scale parallel displacement faults are often characterized by intense fracture zone, dense shear zone, breccia zone and ultra cataclastic zone, and the shear fracture phenomenon is more intense than the other two kinds of faults.

4.3.3 Motions direction determination of the two walls of fault

The activity nature of fault at a certain stage often has relative stability, this kind of movement always leaves certain traces on the fault or its two walls, these traces or attendant phenomena become the main basis for analyzing and judging the relative motion of the walls. But at the same time, the fault movement is complex and a fault often undergoes several pulse-types sliding. Therefore, the complex variability should be fully considered in the analysis and determination of the relative motion of the two walls.

(1) New and old relations of strata

The analysis of the relative new and old of the two walls is helpful to judge the relative motion of the two walls. When the strike direction of the fault is roughly parallel to the strike direction of the strata, namely, it is strike fault, the upthrown wall usually exposures to the old rock strata, or the exposed wall in old strata is the upthrown wall. However, if the strata overturn or the inclination angle of the fault is less than the inclination angle of the strata, the exposed wall in old rock strata is a downthrown wall. If the formation deformation of the two walls is complex, it is a set of strongly compacted folds, so it is not possible to determine the relative motion direction simply according to the new age of two direct contact strata. When the fault and the fold axis are perpendicular or oblique, that is, the transverse fault cuts through the folds, for the anticline, the core of upthrown wall is widened and the core of the downthrown wall is narrowed; for the syncline, it is opposite.

(2) Traction structure

The rock strata that is adjacent to the fault walls often results in a significant arc bending and forms traction folds, and the raised direction of the arc curve indicates the movement direction of this wall. In general, the more intense the deformation, the more tightly the traction folds close. In order to judge accurately, it should be observed at the same time on the plane and section and combine with other characteristics of the relative motion of the fault to make the appropriate conclusion of the relative motion of the two walls.

(3) Scratch and step

Scratch and step are the traces on the fault plane due to friction under the relative dislocation of the two walls. The scratch is shown as a relatively homogeneous set of parallel fine lines. In hard and brittle rock, the rubbing surface is often polished and recrystallization, sometimes with iron, silica and carbonate thin film, as smooth as mirror, and thus become a polished surface. Scratches sometimes appear as thick and deep at one end, thin and shallow at the other. The direction of movement of the wall is generally indicated by the thick and deep end to the thin and shallow end. If you use your fingers to gently touch the scratches, you can feel a smoother along one direction and coarser along the opposite direction. The smooth feeling direction indicates the movement direction of the opposite wall.

On the sliding surface of the fault, there often are some steep scarplets which intersect with the scratch, this kind of fine steep scarplet is called step, and the steep scarplet of the step is generally oriented to the motion direction of the opposite wall.

(4) Feather joints

In the course of the relative movement of the two walls, the tension joints and shear joints in pinnate arrangement are often produced in one or two walls of the fault. These derived joints are oblique to the main fault, and the intersection angles are different due to the mechanical properties. The angle between the pinnate tension joint and the main fault is often 45°, which indicates the motion direction of its wall.

In addition to the pinnate tension joints, the fault-derived joints may have two groups of shearing joints, one with a small angle of intersection with the fault plane, and the intersection angle is generally less than 15°, that is, half of the internal friction angles, and the other group have a large angle intersecting or orthogonal to the fault. A group of joints with small intersection angles indicate the motion direction of its wall. Fault-derived two groups of shear joints are more unstable, or damaged by shear dislocation of two walls, so it is not easy to judge the relative motion of two walls.

(5) Small folds on both sides of the fault

Due to the relative dislocation of the two walls, the strata on both sides of the fault sometimes form complex closed small folds. The axial surfaces of these small folds are often at a small angle with the main fault, and the acute angle indicates the movement direction of the opposite wall.

(6) Fault breccias

If the fault cuts and breaks some an iconic rock stratum, the relative displacement direction of the two walls can be inferred according to the distribution of the angular gravel on the fault plane.

The properties of faults can be determined according to the attitude of the fault surface and the relative sliding direction of two walls.

4.3.4 Scale observation of faults

After the nature of the fault is determined, the extension of the fault is traced, the fault is delineated on the geologic map and the size of the distance is determined.

There are many methods to measure the fault displacement, the simplest method is using the section method to calculate the fault displacement in the outcrop, estimating according to the missing or repetitive formation thickness caused by the fault, or determining the shortest displacement distance according to the distance between the trailing edge of the fenestra and the farthest klippe.

4.4 Field survey and statistics of joints

Joint surveying, there are a variety of different schemata, joint rosette is a more commonly used one, can use the strike direction of the joint preparation, can also use the dip direction of the joint, the preparation method is as follows.

4.4.1 Strike direction rosette of Joint

In an arbitrary radius of the semicircle on the scale network, the joints can be measured according to the strike direction of each 5° or 10° group, and according to statistics in each group of the number of joints the average strike direction angle can be calculated. Straight line and ray are drawn from the center along of the radius, the direction of the ray represents the average strike direction of each group of joints, the length of the ray represents the number of each group of joints, and then use the line to connect the ends of the ray, that is, to get the strike direction rosette of joint such as Figure 4-4.

Each "rose petal" in the rosette represents the strike direction of a group of joints, the length of which represents the number of joints in this direction, the longer the rose petalis, the more the joints are distributed in this direction. As can be seen from the strike di-

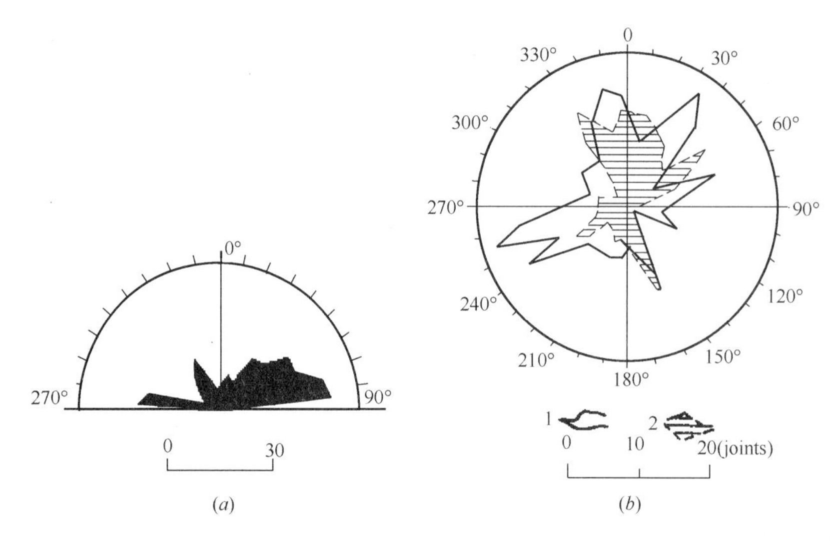

Figure 4-4 Joint rosette

(*a*) strike direction rosette of joint; (*b*) dip direction rosette of joint

1—dip direction of the joint; 2—rosette

rection rosette of joint, the comparative development of the joints are: strike direction angle 85°, 75°, 50°, 335°, 275° five groups.

4.4.2 Dip direction rosset of joint

The measured joints were grouped by 5° or 10° in each group according to the dip direction, and the number of sections in each group was counted, and the average dip direction angle was calculated. Using the method of drawing strike direction rosette of joints, the corresponding points of each group are set out according to the average inclination and the number of joints in the circle with Azimuth, and the points are connected by a polyline, which means that the dip direction rosette of joint, as shown in Figure 4-4 (*b*).

If the length of the radius direction is expressed by the average dip direction angle, the dip direction angle rosette of the joint can also be prepared.

Joint rosette preparation is simple, but the biggest drawback is that it is not in the same picture on the strike direction of the joint, dip direction, dip angle at the same time to express. Tables of statistical survey records of the joint and statistical table of joints are shown in Table 4-1 and Table 4-2.

Statistical survey records of the joint **Table 4-1**

Number	Strike direction	Dip direction	Dip angle	Remark	Number	Strike direction	Dip direction	Dip angle	Remark

Statistical table of joints **Table 4-2**

Directional interval	Average dip direction	Average dip angle	Number of joints	Directional interval	Average dip direction	Average dip angle	Number of joints

4.5 Practice content

(1) Observing syncline and anticline in the area of geological practice, and carrying out the attitude measurement.

(2) Observing the types of faults in the area of geological practice, and carrying out the attitude measurement.

(3) According to statistics of the joints of a certain area, making the strike direction rosette of joints.

Chapter 5 Topographic map and geological map

5.1 Topographic map

The topographic map is the plane map of the topography and ground features. The actual topography and ground features are measured by measuring instruments and is drawn in the topographic map by a specific method according to a certain scale. The topographic map reflects the topography and the actual location of the ground features. General topographic maps are composed of contour lines and ground features symbols, contours lines on the topographic maps represent the fluctuation of topography, the specific symbols represent the ground features.

The topographic map is one of the indispensable tools for field geological work, which is of great significance to the field geological work. With the aid of topographic maps, a preliminary understanding of the topography, ground features and physical geography of a region can be obtained. Some geological conditions can even be analyzed and judged initially. The topographic maps are also helpful for geologic workers to select their working routes and make work plans. In addition, the topographic map is the base map of the geologic map, and the geological workers need to depict the geologic map on the topographic map. The geologic map without the topographic map is an incomplete geologic map, which can not provide a complete and clear concept of the geological structure. Therefore, before the start of field geological work, the geological workers need to be able to read and use the topographic map.

5.1.1 Scale of topographic map

The scale of topographic map is the ratio of the actual distance on the ground to the distance on the map. It is usually marked below the name of topographic map or below the map frame. Commonly used scales include digital scale, linear scale and natural scale.

(1) Digital scale: expressed as a fraction, the numerator is 1, and the denominator is the multiple that is reduced on the graph. For example, 1 : 10000 is one-ten thousandths.

(2) Liner scale: also known as the graph scale, marked with a basic unit length represented by the field distance.

(3) Natural scale: Directly marked the equivalent actual distance to 1cm on the map, such as 1cm = 500m.

The minimum length that one can generally discern in the map is 1 mm, so the horizontal distance of actual ground which corresponding to 1 mm on the map is called the accuracy of the scale. For example, a scale of 1 : 1000, 1mm on the map represent 1m on the actual field, so the accuracy of topographic map with a scale of 1 : 1000 is 1m. The topographic maps with different scales reflect different topographic accuracy; the bigger scale reflects the more accurate topographic features.

5.1.2 Contour lines

Terrain generally indicated by contour lines. The contour line is the connection of the same height adjacent point on the ground, contours are as follows:

(1)The height of one contour line is constant, that is, points on one contour line have the same height.

(2)Self-enclosed, the contour lines will be closed into a curve on their own, if not be closed in this map, limited by the amplitude of the map, it will be closed in the adjacent map.

(3)No cross, no connect, that is, a contour line can not be bifurcated into two and two contour lines can not be combined into one, except cliffs.

Contour lines reflect the basic information of topographic relief; the topographic map is the horizontal projection map of contour lines. The mean sea level (MSL) of the Yellow Sea is the starting point for calculating the elevation, that is, the zero point of the contour lines. According to the zero point, the absolute elevation of any terrain can be calculated.

The contour interval is the vertical distance between two adjacent imaginary horizontal sections which cutting the terrain. In a topographic map with certain scale, contour interval is fixed.

The contour horizontal distance is the horizontal distance between adjacent contour lines on the topographic map. The length is related to the terrain. The terrain is steep and the length is short and the converse is otherwise long.

5.1.3 Features of all kind of terrain

5.1.3.1 Hill and depression

As shown in Figure 5-1, the hill and depression are closed curves, and the terrain can be distinguished according to the elevation. Among the closed contour lines, the higher contour line appear in thc inner ring is the hill, as shown in point A in Figure 5-1; the lower contour line appear in the inner ring is the depression, as shown in B point in Figure 5-1.

The saddle between adjacent hills, on the topographic map, is the adjacent group of contours lines which represent the hill, with the saddle located in the middle, as shown in point C in Figure 5-1.

The watershed located in the middle of adjacent depressions. In the topographic map, the middle of adjacent groups of closed contours line representing the depression is watershed, as shown in point D in Figure5-1.

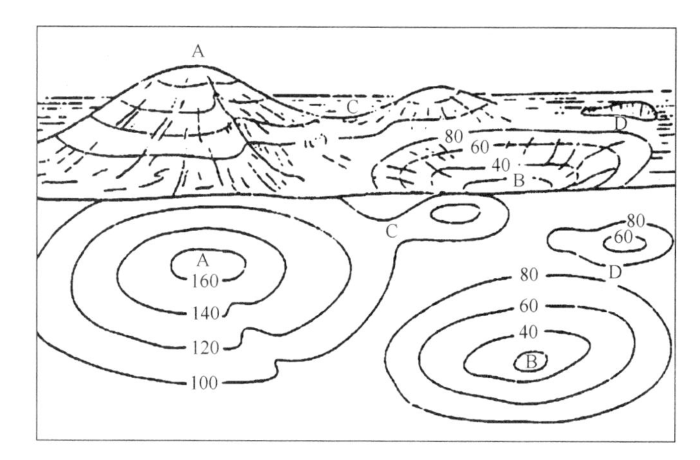

Figure 5-1 The contour line of hill and depression

5.1.3.2 Mountain slope

The section of mountain slope generally has four types: uniform slope, convex slope, concave slope and stepped slope, of which the density of contour line interval is different.

(1) Uniform slope: the horizontal distance between adjacent contour lines is equal.

(2) Convex slope: the contour line interval is mass in the upper, and thin in the bottom.

(3) Concave slope: the contour line interval is thin in the upper, and mass in the bottom.

(4) Stepped slope: the density of contour line is the same, but horizontal distance is different in upper and bottom.

5.1.3.3 Cliffs

When the slope is steep, namely it is a cliff, the contour lines overlap into a thick line, or the contour lines intersect, but the intersection point must appear in pairs. And some special symbols can add in the overlap of contour lines.

5.1.3.4 Ridge and valley

As shown in Figure 5-2, the contour lines of valleys and ridges have almost the shape. Therefore, they are differentiated according to the elevation of the contour lines, indicating that the contour lines of ridges are convex to the lower, as shown in the point A in Figure 5-2; and that the contour lines of valley are convex to the bottom, as shown in the point B in Figure 5-2.

5.1.3.5 River

As the contour lines pass through the river, they can not cross the river vertically and must travel upstream along the banks and then across the river, back again downstream to leave the riverbank, as shown in Figure 5-3.

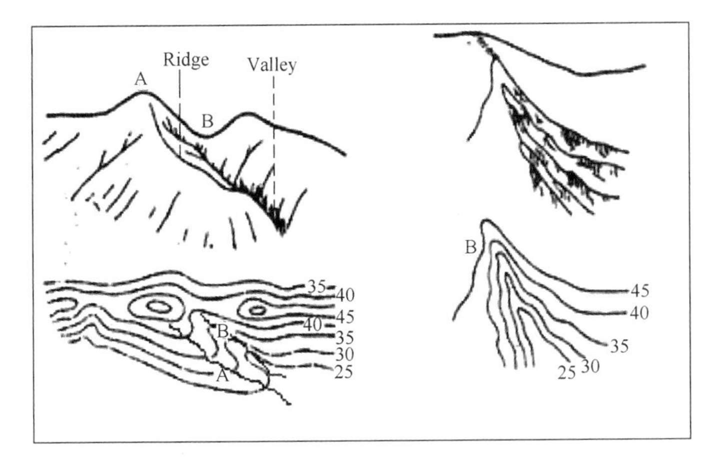

Figure 5-2 The contour lines of ridge and valley

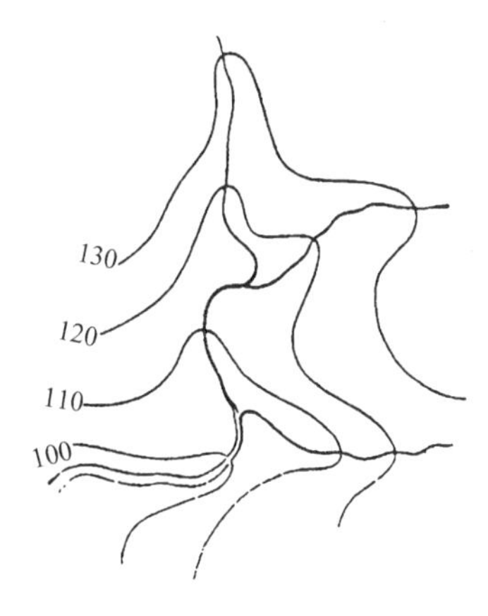

Figure 5-3 Contour lines of river

5.1.4 Ground features symbols

In topographic maps, different symbols represent a variety of ground features; there are three commonly used symbols.

(1) Proportional symbol is a similar figure, which is also an outline symbol, which is drawn on the graph in kind according to the scale of the topographic map.

(2) Non-proportional symbols, when the actual area of the ground feature is so small that it can not be drawn on the graph with the scale of the topographic map, its position is usually marked with a specific symbol on the topographic map.

(3) Linear symbols, the length is proportionate, and the width is not disproportionate, when the ground features was like-strip or elongated shape such as railways, roads, and the length can be drawn according to the scale, but the width can not.

5.1.5 How to read the topographic map

The purpose of reading a topographic map is to understand and be familiar with the terrain of a certain area including the understanding of the various elements and their relationships of the terrain and ground features, not only to understand the ground features such as mountains, river, villages, roads and topography, but also to analyze the topographic map, put the various symbols and markers together to serve the geological work. The steps to read a topographic map are as follows:

(1) Read the name of the geological map. The map usually named after the most important place in this topographic map, the area of this topographic map can be roughly determined according to the map name.

(2) Recognize the direction. Except for some maps, the direction of the topographic maps is north in the upside and south in the downside, left is west and right is east. Some

topographic maps have latitude and longitude, and the latitude and longitude can be used to determine the direction.

(3) Confirm the location. From the latitude and longitude marked on the frame understand the location of the topographic map.

(4) Understand the scale. Understand the area of map, the precision of the topographic map and the distance of contour lines according to the scale.

(5) Identify the terrain. According to the characteristics of the contour lines, identify the distribution and characteristics of mountains, hills, plains, mountain peak, valleys, gentle slopes, steep slopes, cliffs and other terrains in this map.

(6) Understand the ground features. Combine with the legend to understand the location of the region's features such as rivers, lakes, settlements and other distribution, so as to understand the region's geography, economy and culture. For example, Figure 5-4 shows the topographic map of an area.

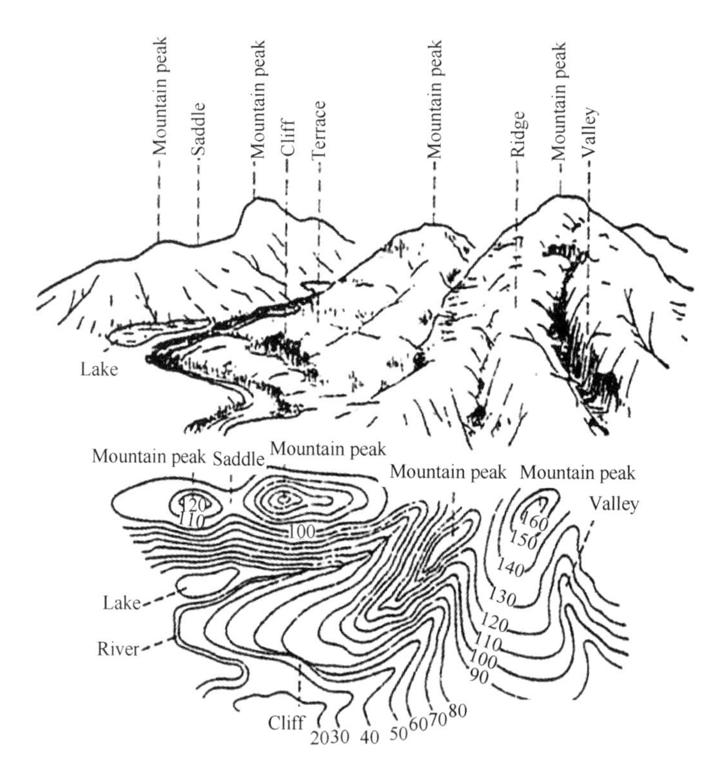

Figure 5-4 Integrated topographic sketches and topographic map

5.2 Application of topographic map in geological work

5.2.1 Make terrain profile

Terrain profile is based on hypothetical vertical plane cuting the terrain and geting the cross-section. The cross-section line is the intersection line of the ground surface and the cross-section. Geological workers need to make terrain profiles, because putting the geological profile and terrain profiles together, the geological phenomena and spatial connection can be easier to reflect. Terrain profiles can be produced based not only on the topographic map, but also on the field map.

5.2.1.1 Make the terrain profiles based on topographic map

(1) Select the location of the desired topography on the topographic map, as shown in Figure 5-5, and draw the section line AB.

(2) Make the baseline. Draw a straight line at the middle and lower part of the graph paper as the baseline ab, and sets the baseline altitude at 0.00, which may also be the value of the lowest height of contour line on the section line, as shown in Figure 5-5, where the baseline altitude is 50m.

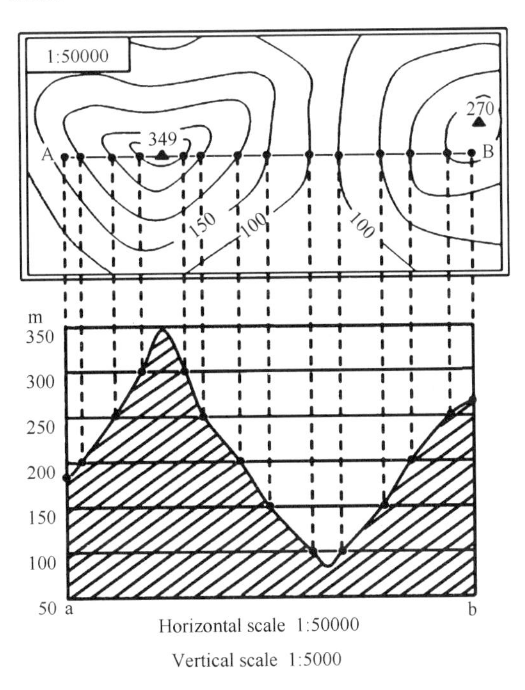

Figure 5-5 Making a terrain profile

(3) Make a vertical scale. Make a vertical line on the left of the baseline. Generally, if the vertical scale and the topographic scale are consistent, then the topography profile should be consistent with the actual one. If the terrain relief is very gentle, for special needs, enlarge the vertical scale to make the terrain changes more obvious.

(4) Make vertical projection. The baseline ab is parallel to the section line AB, and the points of contour line on the topographic map intersecting with the AB line are vertically projected onto the respective elevation above the baseline ab to obtain the corresponding topographical points. The general direction of the profile line is the north or west on the left, the east or south on the right.

(5) Get a curve. Use the smooth curve to connect the obtained topographic points and get the contour lines of the terrain.

(6) Mark the location of objects, the name of the map, the scale and the direction of the cross section, and make the decoration for beauty.

5.2.1.2 Make the terrain profiles based on field mapping

When conducting the geological survey in teaching practice, it is often required that a topographic profile be drawn on site to reflect the geology on the route of geological survey. It is needed to determine the location of the section, the direction of the section, the length of the section and the section scale according to theprecision requirements. The drawing steps are similar to the previous method, except that the horizontal distance and elevation difference are determined by observing on the spot. Therefore, it is the key point to determine the horizontal distance and elevation to make a good topographic profile. When the cross section is short in length, the horizontal distance and elevation difference can be roughly measured or measured by walk. When the cross section is long in length, the horizontal distance and elevation difference can only be calculated by eye estimation or refering the topographic map or calculating according to the barometer. Usually, we sketch the profile part by part. Sketch one part after observing one part; otherwise it is easy to distortion. If the terrain is simple and the worker is skillful, it also can be one shot.

5.2.2 Set points in field with topographic map

When carrying out geological work in the field, it is often necessary to accurately mark some observation points on topographic maps, such as geological points, ore points, work points and hydrological points. These points are fixed points in the regional geological survey work. There are two ways that how to use topographic map to set points.

(1) When the precision requirement is not high, such as small-scale mapping or preliminary survey, visual estimation method can be used for sets points, that is, according to the measured point around the terrain, the relationship between the distance and direction of ground features, with the eyes to determine the location of measuring point on the topographic map.

When set points with visual estimation method, the first step is to use the compass to set the orientation of the topographic map at the observation point, the long edge of compass alongside the east or west of the frame of topographic map, move the topography and compass together, so that the North for the quasi-dial 0 °. After that, the due north direction of the frame is coincident with the due north direction of the observation point, the direction of topographic map is coincident with the real direction in field. At this time, some linear objects such as rivers and roads should parallel to the river or road marked on the topographic map.

When the direction of topographic map is fixed, search and observe the characteristic and obvious ground features or topography and estimate their relative position (such as direction, distance, etc.) with the observation point. Then locate the observation point and mark it on the topographic map according to the relationship.

(2) In the geological survey with a larger scale, if the required precision is high, the intersection method is required.

Firstly, determine the direction of topographic map according to the visual estimation method, then find out three known points that are not on a straight line and have been marked on the topographic map near the observation point, such as triangle points, mountain peak, buildings, etc. and measure the relative orientation of the observation points to these points with the compass. At this moment, the compass sighting board faces the observer, raises the little hole of the sighting board, and then uses the hole and the centerline of the retroreflector to aim at the selected triangle point or mountain peak. When the three points are in one line and the blisters are centered, read out the value of the compass as the orientation of the survey line, that is, the relative orientation of observation points and the known point, record the orientation of three survey lines.

Find out three known points on the map and plot these points with protractor. Get three lines passing through three known points on the topographicmap, the intersection point of the three lines is the position of the measure point. If the three lines cross into a triangle (called an error triangle) due to the measurement error, instead of one point, the position of measure point should take the small point of triangle.

Pay attention to two points when apply to field work.

(1) When measuring the direction of the line, if the sighting board of the compass is aimed at the known point, the value of south arrow of compass is the orientation of the observed point. The value of north arrow of compass is the direction of the known point at this observation point. In order to avoid confusion, record the value of north arrow of compass when the sighting board of the compass is faced the unknown point (the direction of the point).

(2) When using the protractor to draw the direction of measured line on the map, pay attention to the geographical coordinates instead of the position on the compass.

In field work, the visual estimation method and the intersection method is mixed to correct each other so that the observation points can be positioned more accurately. For example, after the error triangles are drawn by the three-point intersection method, using the visual estimation method to find out the location and elevation of special features to revise the position of measuring points.

5.3 Geologic maps

Geologic maps reflect geological phenomena and geological conditions in a certain region with prescribed legends, symbols and colors. It is based on geological data measured in the field and are projected on the base map at a certain scale. The geologic map is one of the important results in geological surveys. The planning, design and construction phases of the project construction are based on the geological survey data. The geologic map is a compilation of various survey data and is an important map data that can be directly used for production.

5.3.1 Types of geologic map

There are many types of geologic maps. Due to the different purposes of construction, the geological contents of the geologic maps also have their own special emphasis. The geologic maps commonly used in several projects are described below.

5.3.1.1 General geologic map

General geologic map, also known as terrain and geologic map, is the basic map of the terrain, lithology and geological structure conditions of a certain area. General geologic map projects geological boundary lines such as stratigraphic boundaries of different geologic times and major structure lines that are exposed to the surface, including geological plans, geological cross section and comprehensive strata log diagram.

(1) The geological plans indicate the various geological data obtained in the field such as geomorphology, stratigraphy, geological structure, natural geological effects, hydrogeological conditions by using a variety of legends.

(2) The geological cross section, reflects the deep stratum and geologic structure.

(3) The comprehensive strata log diagram is a columnar cross-sectional diagram that synthetically reflects the stratigraphic sequence, thickness, lithology characteristics and regional geological development history in the surveying area according to a certain scale and legends, as shown in Table 5-1.

The comprehensive strata log diagram (plotting scale 1 : 1000) Unite: mm Table 5-1

Geological chronostratigraphic			The lithostratigraphic		Code number	Lamination number	Histogram
Erathem	System	Series	Formation	Member			
10	10	10	15	10	10	10	30

Layer thickness (m)	Segment thickness (m)	Lithology description and ore-bearing	Paleontological characteristics	Sediment	Environment	Specimen	Remark
15	15	100	60	30	15	15	20

5.3.1.2 The engineering geologic map

The engineering geologic map show the spatial distribution and relationship of engineering geological conditions in a certain area or building area according to a scale, and it is a geologic map that combines the indicators of geological engineering construction, mapping, compilation.

5.3.1.3 The quaternary geological map

The quaternary geological map, which using different colors, patterns and symbols, map the quaternary sediments genetic type, quaternary strata, quaternary volcanic lithology and age, as well as the quaternary geological structure phenomenon of a certain area in the topographic map at a certain scale, is called the general quaternary geological map or regional quaternary geological map.

5.3.1.4 The hydrogeological map

The hydrogeological maps refer to the geological maps that reflect the distribution, burial, formation, transformation and dynamic characteristics of groundwater in a certain area. The maps mainly show the groundwater types, occurrences, properties and distribution, they are the main form of research results of hydrogeological surveys in a certain area.

The above geological maps should include the name, legends, scale, direction and responsibility list, etc. , in which the illustrations strictly require from top to bottom or from left to right, the strata are arranged from new to old, first is strata and magmatic rocks, then geological structure and so on.

5.3.2 The scale of geologic map

The scale is an index that reflects the accuracy of the map. The larger the scale is, the higher the precision of the map is and the more detailed and accurate the content reflected. Geologic maps can be divided into: small-scale geologic maps (less than 1 : 200000 to 1 : 1000000); medium-scale geologic maps (1 : 50000 to 1 : 100000); large-scale geologic maps (greater than 1 : 1000~1 : 25000).

5.3.3 Representation of geological structure

5.3.3.1 Horizontal stratum

Topographic contour lines are the closed curves connected by adjacent points on the ground with equal elevations. The boundary of the strata in the horizontal strata is parallel or coincident with the topographic contour lines, as shown in Figure 5-6.

5.3.3.2 Tilted stratum

The boundary of the tilted stratum is a curve in the geologic map that intersects with the topographic contour lines as "V" or "U". The "V" shapc on the geologic map is also different, as shown in Figure 5-7.

(1) When the dip direction of the rock stratum or geological interface is opposite to the slope of the ground, the direction of the outcrop of the rock stratum or the geological interface is consistent with the topographic contour lines. In the valley, the tip of the "V" shape points to the upper stream of the valley.

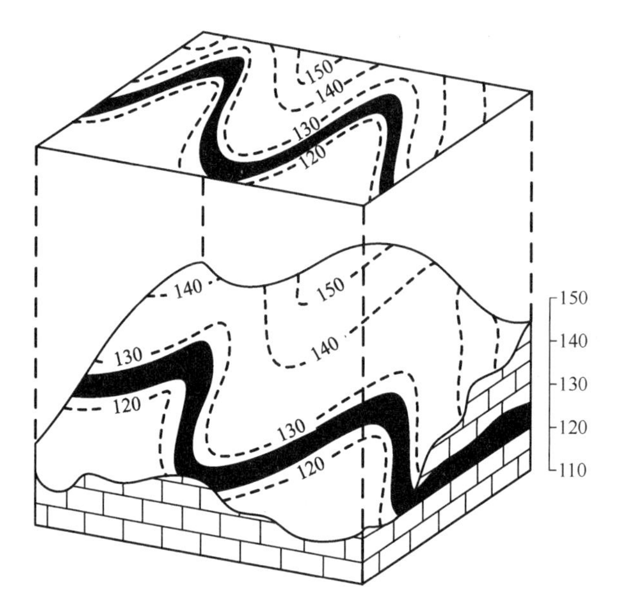

Figure 5-6 Horizontal stratum

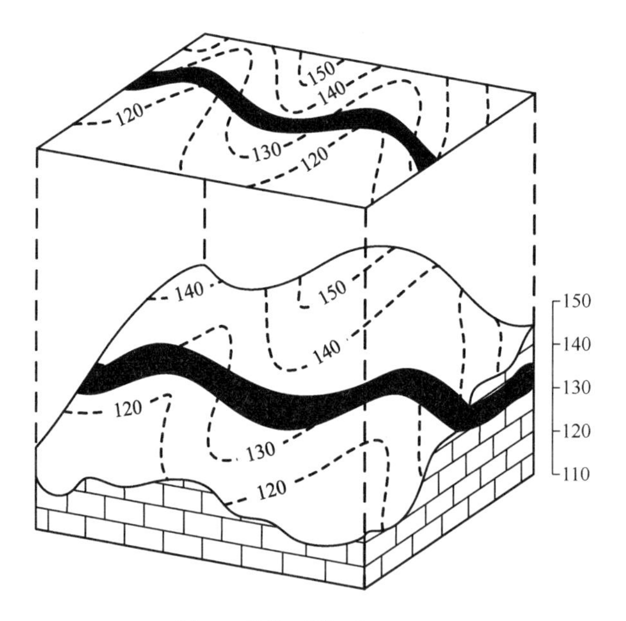

Figure 5-7 Tilted stratum

(2) When the dip direction of the rock stratum or geological interface is consistent with the slope of the ground, if the inclination angle of the rock stratum is greater than the slope of the ground, the bending direction of the rock stratum or the outcrop line of the geological interface is opposite to the bending direction of the topographic contour lines. The outcrop of the rock stratum or geological interface forms a "V" shape in the valley with its tip pointing downstream.

(3) When the dip direction of the rock stratum or geological interface is consistent

with the slope of the ground, if the inclination angle of the rock formation is less than the slope of the ground, the bending of the rock stratum or the outcrop of the geological interface is similar to the curvature of the topographic contour lines. The outcrop of the rock forms a "V" shape in the valley that points to the upstream.

The rock stratum can also be identified by legends, such as the symbol "⅄". The long line indicates the strike direction of rock stratum and the short line indicates the dip direction.

5.3.3.3 The vertical stratum

The boundaries of vertical stratum are not affected by the topographic contour lines and extend straight along the strike direction of rock stratum, as shown in Figure 5-8.

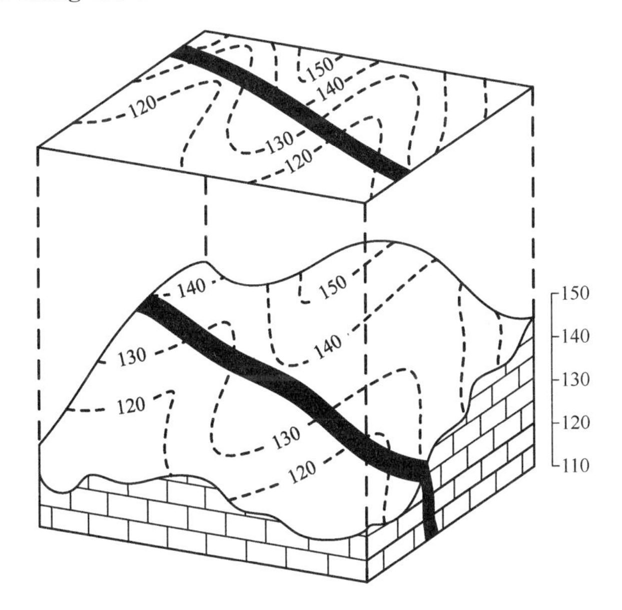

Figure 5-8 Vertical stratum

5.3.3.4 The fold

The fold is generally identified according to the legends (Figure 5-9). If there are no legends, it is needed to be identified according to the symmetry distribution between the new and old rock formations.

Figure 5-9 The legends of fold

(*a*) Syncline; (*b*) Anticline

5.3.3.5 The fault

The faults are identified according to the legends, as shown in Figure 5-10. If there are no legends, the faults are identified based on the distribution of rock formations such as repetition, absence, interruption, width variation and misalignment.

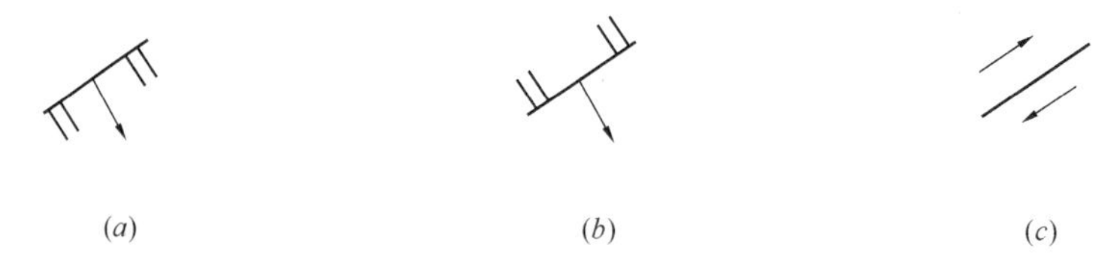

Figure 5-10 The legends of fault

(*a*) Normal Fault; (*b*) Reverse Fault; (*c*) Parallel displacement fault

5.3.3.6 Contact relationship of strata

The expression of conformable contact and parallel unconformity on the geologic map is consistent with the bending features of boundary of adjacent rock formations. The age of former rock formations is continuous, and the age of latter is discontinuous. The feature of the angular unconformity on the geologic map is that the boundary of the new rock formations blocks the boundary of the old rock formations. The intrusive contact breaks the boundary of the sedimentary rock where the intrusion body is exposed but does not move on both sides of the intrusion body; the sedimentary contact shows that the intrusion body is covered and interrupted by the sedimentary rock formations.

5.3.4 Basic concept of geologic map

A formal geologic map should have names, scales, legends and responsibility tables (including group and person of drawing map, drawing date of map, source of data, etc.).

(1) Name of map

The map of name indicates the type of map and the area where the map is located. It is generally named after the main towns, residential areas or major mountains, rivers, etc. For example, if the scale is large and the area of the map is small, the place name is not well-known or the name is duplication, the name of the province, city or county which it belongs to should be written in the place name such as the geologic map of Haidian District of Beijing. The name of map should be written in an upright and beautiful font and in the middle of the map frame or in an appropriate position in the map.

(2) Scales

The scale indicates the detailed degree of actual geological conditions. The scale of a geologic map is the same as that of a topographic map, with a digital scale and a line scale. The scale is generally placed at the top outside the frame, under the name of the map or in the middle position below the map.

(3) Legends

The legend is an indispensable part of a geologic map. Different types of geologic maps have their own legends of geological content. The legends of common geologic maps use various prescribed colors and symbols to indicate the age and nature of the formation and rock. The legend is usually placed on the right or bottom outside the frame, or it can be placed in the frame where enough to arrange the space for the legend. The legends should be arranged in a certain order, generally in the order of stratum, rock and structure, with the words "legends" in front of them.

① The arrangement of stratigraphic legends is from top to bottom and from new to old. If placed below the map, it is generally arranged from left to right and from new to old. The legends are all drawn into rectangular grids of appropriate size and arranged in neat rows. The colors and symbols marked in the grid are the same as the colors and symbols in the same layer on the geologic map, and the stratigraphic age and major lithology are indicated in appropriate positions outside the grids. The extrusive rocks and metamorphic rocks which the era is identified must be arranged in the corresponding positions according to their ages. The legends of magmatic rock are placed after the stratigraphic legend, the rock masses which the age is identified should be ranked from the old and new ages, and the rock masses in the unconfirmed ages should be arranged in the order of acidity to basicity.

② The legends of geological structure symbols are placed behind the legends of the strata and rocks. The general order: geological boundaries, fold axes (only in the geological structure maps), faults, joints, bedding, cleavage, schistosity, flow lines, flow surfaces and lineation occurrence factors. In addition to the fault line is red lines, the rest legends are black lines.

③ The strata, rocks, geological structures and other geological phenomena that exist in the map all should have relevant legends. No legends if the symbols do not exist in the map. The legends of topographic maps are not marked on the geologic map.

④ Responsibility tables

The drawing date of map is on the upper right side of the frame; the mapping group, the technical director and the editor are on the lower left side; the organization, the compiler and the preparation date of the cited data are on the lower right side. You can also list the above content as "responsibility form" on the lower right outside the frame.

Longitude and latitude are drawn on the small-scale map to indicate its geographical location. If the geologic map is one of the tessellations of the international topographic map, the name of map and framing number should be used in the same way as the topographic map.

5.3.5 How to read the geologic map

(1) First, read the name, scale and area so that have an overall concept for the area in the geologic map. Then, read the location and direction of the map. Generally, the north arrow is used as the basis. If there is no direction, the north is the top or the direction is

determined according to the coordinates.

(2) Read the legends. The legends on the plan map, section map and histograms map are all consistent. The legends of the geologic map are drawn on the right side of the frame, and all the strata symbols, geological structure symbols, etc. in the map are listed from top to bottom in chronological order from new to old. At this point, it should be noted whether there is a lost strata.

(3) Analyze the landform. Through the distribution characteristics of contour lines and river systems, we can understand the situation of mountains and rivers in the area, the ups and downs of the terrain, the topography and morphological characteristics.

(4) Understand the distribution of stratigraphic and the lithology. With reference to the legends, read the distribution and status of the stratum, the relationship between the old and new stratum, the relationship with the terrain.

(5) Geological structure type. Through the legends, understand the distribution of geological structures such as type, size and distribution of faults and folds, the strike direction of tectonic lines in the area, the relationship between terrains, etc.

(6) Magmatic rock. If magmatic rocks are exposed in the area, the active age and the intrusion or eruption of magma should be clarified, and then the occurrence of magmatic rock should be determined according to the occurrence and morphological characteristics of magmatic bodies.

(7) Evaluation. An initial evaluation for the stability of the building site should be done according to the geological conditions in the map.

Chapter 6 Some comparisons of engineering geology at home and abroad

6.1 Introduction to some building code systems

6.1.1 Introduction to European building code system

In March 1961, 13 countries of the European Union, including Britain, France, Germany and Italy, met in Paris and established the European Committee for Standardization (referred to as the CEN for short). CEN foused on promoting Standardized collaboration among member states, and development European standards for the European Union (except the field of electricians and telecommunications).

Now, there are 30 members participating in the development of European Code, including Austria, Belgium, Bulgaria, Cyprus, the Czech Republic, Denmark, Estonia, Finland, France, Germany, Greece, Hungary, Iceland, Ireland, Italy, Latvia, Lithuania, Luxembourg, Malta, Netherlands, Norway, Poland, Portugal, Romania, Slovakia, Slovenia, Spain, Sweden, Switzerland, UK (refer to website: http: / /www. cen. eu). European code is published in three official languages, namely English, French and German.

In 1975, Commission of the European Community in accordance with treaty 95, developed technical codes in the field of construction and produced the first European code in the 1980s. In 1989, the Commission of the European Community gave the task of formulating and publishing European codes to the European Committee for Standardization (CEN). After decades of development, the CEN 250 technical committee (referred to as CEN/TC-250 for short) is now responsible for all construction codes.

The 30 member countries of the CEN have always adopted the European code as their national standard, and have added Annexe according to their national conditions to define their own design codes. Therefore, each member country of CEN has its own national codes in accordance with its own actual situation. For example, The British has BS EN 1998-1:2005 to correspond to the first part of Eurocode 8 (EN 1998-1 : 2004), German has DIN EN 1998-1 : 2006.

The current European structural codes are mainly composed of ten codes from Eurocode to Eurocode 9, as shown in Table 6-1. Codes involved in Engineering Geology mainly include Eurocode 7 (geotechnical design code) and Eurocode 8 (Design of structures for earthquake resistance code).

European codes **Table 6-1**

Code		Name
Eurocode	EN 1990	Basis of Structural Design
Eurocode 1	EN 1991	Actions on Structure
Eurocode 2	EN 1992	Design of Concrete Structures
Eurocode 3	EN 1993	Design of Steel Structures
Eurocode 4	EN 1994	Design of Composite Steel and Concrete Structures
Eurocode 5	EN 1995	Design of Timber Structures
Eurocode 6	EN 1996	Design of Masonry Structures
Eurocode 7	EN 1997	Geotechnical Design
Eurocode 8	EN 1998	Design of Structures for Earthquake Resistance
Eurocode 9	EN 1999	Design of Aluminium Strucutres

6.1.2 Introduction to American building code system

The United States is a federal country, so its building code system is unique comparing to other countries in the world. The formulation of building codes is the responsibility of the state governments. Each state is responsible for the legislation of building safety. The state, city and county governments promulgate and implement building technology codes.

Most of the work on the model code is done by associations or standards organizations, most of which are independent nonprofit or private organizations, not managed by government agencies or organizations. According to the provisions of the constitution of the United States, states have the right to legislate and decide to adopt the model codes of any association as the technical codes of their states according to their state's conditions. Therefore, there are great differences between states in whether to adopt the existing model codes and how to adopt the model codes.

In the first half of the 20th century, three nonprofit and private organizations, namely BOCA (1915), ICBO (1922) and SBCCI (1941) have started to seek and formulate standardized model codes.

Model codes referred to a number of standards developed with associations or societies such as the American Society for Testing and Materials (ASTM), the American Society of Civil Engineers (ASCE), the American Concrete Institute (ACI).

In 1972, the board of directors of the three agencies (BOCA, ICBO and SBCCI) formed the Coucil of American Building Officials (CABO), but not in place of the three agencies. The purpose of the CABO is to strengthen the status and role of model codes and demonstrate policy consistency by consultation.

In 1994, the establishment of the International Code Council (ICC) ended the long-term division of the American building model code. One of its tasks was to formulate a unified national model code, which was based on the three sets of model code of the international Building Officials and Code Administrators (BOCA).

The ICC is committed to developing national building safety, fire protection and energy efficiency codes for residential, school and commercial buildings, thus changing the situation that different regions in the United States adopt different codes.

In 1997, ICC issued the International Building Code (Draft), IBC. Since then, the model codes formulated by BOCA, ICBO and SBCCI were not updated and merged into a unified IBC code, which was officially published in 2000. So far, ICC has published 15 model codes as a series of I-code as shown in Table 6-2. ICC code is a unified, unrestricted, regionally recognized national building code in the United States. Although some states are still developing their own codes, most states, cities and counties in the United States have chosen to adopt international codes and building safety codes established by ICC. These codes also serve as a basis for federal real estate to build buildings outside the United States. Many countries around the world also refer to these codes.

ICC codes **Table 6-2**

NO.	Name
1	International Building Code
2	International Plumbing Code
3	International Energy Conservation Code
4	International Private Sewage Disposal Code
5	International Existing Building Code
6	International Property Maintenance Code
7	International Fire Code
8	International Residential Code
9	International Fuel Gas Code
10	International Swimming Pool and Spa Code
11	International Green Construction Code
12	International Wildland Urban Interface Code
13	International Mechanical Code
14	International Zoning Code
15	ICC Performance Code

The U. S. building technology regulatory system is mainly composed of the following three parts, which form a certain hierarchical relationship.

(1) Model code

Model code has the highest level, the highest acceptance, the lowest content, and certain standards are adopted or referenced in each model code.

(2) Consensus standard

Consensus standard is the widely recognized content that has not yet been upgraded to the mode code, or the specific description of the mode code content, which is used to guide the specific engineering design work.

(3) Resource document

Resource document records the content at a deeper level, explaining the principles and background of the content stipulated by the code and standard, and also including the latest research results of each code and standard. Resource document are often quoted by the model code.

The three major parts of the building technology law system are not strictly progressive, but developed and updated at the same time including interpenetration and infiltration.

Building technology regulations refer to supporting technical standards compiled by various professional associations and societies. These professional associations and institutes are mostly non-official, ANSI-authorized Standard Development Organizations (SDO) such as ASCE, ACI, ASTM, etc.

6.1.3 Introduction to Chinese building code system

China's engineering construction standards have gone through a long development process, and gradually formed a construction standard system combining "construction regulations, namely mandatory standards and recommended standards".

According to the order of legal effectiveness from high to low, China's building codes have five main components.

(1) Laws, such as the building law;

(2) Administrative rules and regulations, such as regulations on the administration of construction project survey and design;

(3) Local regulations, such as regulations on the construction market governing in Beijing;

(4) Departmental rules, such as measures for the administration of national standards for project construction;

(5) local government regulations, such as shanxi provincial administrative measures on bidding and tendering for construction projects.

China's engineering construction standard system adopts the construction technology standard system of "two types" (mandatory standards and recommended standards) and "four levels" (national codes, industrial standards, local standards and enterprise standards). National standards are technical requirements that are uniformly implemented throughout the country. Industry standards are technical requirements that need to be uniformly implemented in a certain industry nationwide in the absence of national standards.

Industry standards are technical requirements that need to be uniformly implemented in a certain industry nationwide in the absence of national standards. Local standards are technical requirements that are not found in the former two standards and need to be uniformly implemented within an administrative area. Enterprise standards are technical requirements that are uniformly implemented within an enterprise in the absence of the other three standards. The lower level standards should be abolished after the higher level standards are issued.

Engineering construction standard system published in 2003 makes statistics on the standards of 17 majors including urban and rural planning, urban construction and housing construction. The statistical results are shown in Table 6-3, in which the total number of standards includes current standards, standards in preparation and standards to be prepared.

Engineering construction standard system information statistics **Table 6-3**

No.	Major name	Basic standard	Universal standard	Special standard	Total number
1A	Urban Planning	5	10	32	46
1B	Town Construction	2	6	18	26
2	Reconnaissance	9	13	49	71
3A	Urban Traffic	5	10	9	24
3B	Urban Railway	12	57	200	269
4	Road and Bridge	3	14	35	52
5	Water Supply and Discharge	8	17	55	80
6	Gas	5	11	16	32
7	Heat Supply	5	7	10	22
8	Environment and Health	5	19	35	59
9	Sightseeing and Gardens	8	11	19	38
10	Engineering Anti disaster	13	14	24	51
11A	Architectural Design	9	24	62	95
11B	Electricity Design	2	10	72	84
12	Building Basement	1	2	13	16
13	Building Structure	5	10	42	57
14A	Construction Quality	5	48	50	103
14B	Construction Safety	3	13	23	39
15	Maintenance and Real Estate	5	34	35	74
16	Indoor Environment	6	14	65	85
17	Information Technology Application	10	12	30	52
Total number		126	356	893	1375

Chinese codes involved in engineering geology mainly include codes for seismic design of buildings and codes for geotechnical investigation.

In 1959, China put forward the first draft code for seismic design, which mainly includes buildings, roads, Bridges, DAMS, water supply and drainage, etc.

In 1974, the code for seismic design of industrial and civil buildings, China's first officially approved code for seismic design, was published. In 1978, the code was revised and the code for seismic design of industrial and civil buildings TJ11—78 was published.

In 1982, the academy of architectural sciences was responsible for revising the TJ11—78 code, and in 1990, the code for seismic design of buildings GBJ 11—1989 was formally implemented.

After continuous development, the revised outline of GBJ 11—1989 was adopted in 1997. Through engineering practice and summarizing the experience and lessons at home and abroad, the code for seismic design of buildings GB 50011—2001 was formulated in 2001 and jointly issued by the ministry of construction and General administration of quality supervision, inspection and quarantine.

The 1960s is the initial period of geotechnical engineering standardization in China, which indicates that Chinese geotechnical engineering technology starts from scratch and matures in large-scale engineering construction.

By the early and mid-1970s, a number of specifications for geotechnical engineering had been born and revised, among which the code for geological investigation of industrial and civil construction engineering (TJ 21—77) was an important one, marking the beginning of the establishment of a national standard for geotechnical engineering investigation in China. Fundamental changes to the specification TJ 21—77 began in 1986, were reviewed in 1990, and approved and promulgated in 1994. In 1998, the ministry of construction issued a document to revise the code, and completed the revision in 2001. On January 10, 2002, the ministry of construction approved and promulgated the code for geotechnical engineering investigation (GB 50021—2001), which came into force on March 1, 2002.

6.2 Definition and discrimination of active faults at home and abroad

As a term of structural geology, active fault was proposed by A. C. Lawson (1908), H. Q. Wood (1916), B. Wilis (1923) and Li Siguang (1926) in the early 20th century. Since the 1960s, a series of major earthquakes have taken place in the world, and it has become an accepted fact that the occurrence of earthquakes is often controlled by active faults.

In 1976, the epicentre of the Tangshan earthquake in China was distributed along the NO. IV-V Tangshan seismogenic fault.

The 1987 Whittier Narrows earthquake caused more than $300 million in economic damage, experts said the earthquake was caused by a hidden active fault in the city of Los Angeles and could cause another strong earthquake in the future.

In 1995, the hardest-hit areas of Kobe earthquake in Japan were concentrated along the active fault. More than 90% of the earthquake deaths were concentrated in the width range of 2 ~ 3 km along the active fault.

In 1999, the hardest hit area of the Izmit earthquake in Turkey was on the northern branch of the western part of the north Anatolia fault. All the buildings built on the fault collapsed.

The hardest-hit area of the 1999 chi-chi earthquake in Taiwan was on the chelangpu fault.

The concept of active fault has been controversial since it was put forward. Active fault is first thought to have been active or new faults since the quaternary period (2 ~ 3Ma) or the late tertiary period (about 3Ma) after the alpine-himalayan movement. R. e. Wallace of the United States geological survey believes that active faults that have shifted in the past 10000 years or in the late quaternary can be called active faults. C. r. Allen of the United States says faults that have been active in the last 100000 or 10000 years are called active faults.

In 1956, at the symposium of the first neotectonic movement held by the Chinese academy of sciences, it was proposed that new faults and quaternary faults should be used to describe the faults that led to the faults of Cenozoic or quaternary strata or that had obvious geomorphic features. In 1989, the earthquake disaster prevention department of the state seismological bureau of China stipulated: "active faults are those that have been active during the quaternary period, especially since the late pleistocene (100000 years), and are likely to be active in the future.

Active faults are generally judged by the fact that they are still active, while potential active faults are judged by various criteria. It is indisputable that there have been active recorded faults in human history, but there are different understandings and limitations on the identification of potential active faults in recent geological history. Some people regard the modern geological historical period as the holocene (11000a). Some people put it as recently as within 35000a. Some people put it as within the late pleistocene (recently 100000a or 500000a). Others base their judgement on repeated activity in recent geological historical periods, such as the quaternary period.

In 1973, the U. S. atomic energy commission made three rules on how to determine active faults:

(1) The fault had one more activity in 35000 years.

(2) The fault is associated with capable faults.

(3) The small seismic events or peristalsis has been recorded along the fault.

Subsequently, International Atomic Energy Agency added two new provisions to the definition of the United States Atomic Energy Commission:

(1) The fault has been active since the late quaternary.

(2) The fault has ground crack evidence.

In Europe and America, active faults are mostly studied in combination with major engineering construction. Any fault that meets the above criteria is active fault, orcapable fault, which is used as the basis for judging the stability of the project site.

Based on advanced scientific research methods, Japan takes the occurrence time of faults as the standard and divides active faults into three categories:

(1) Faults that have been active since quaternary (about 2Ma).

(2) Faults that have been active since the holocene (about 11000 years).

(3) Faults that have been active in his history since the holocene (100 ~ 200 years).

Deng Qidong proposed three definitions based on the crustal dynamics process of China and the current research level:

(1) The active fault refers to the active fault that have been active or being active from quaternary (2 ~ 3Ma) to now or since Neogene.

(2) Capable fault refers to the fault that has been active since middle and late pleistocene (100000 ~ 500000 years).

(3) Seismic fault refers to the fault that has the destructive history of seismic records since 1000～2000 years and a strong prehistoric seismic activity since the holocene (110000 years).

The qualitative evaluation of active faults has been unable to meet the site selection and design needs of major projects.

6.3 Classification of site soils at home and abroad

The three code systems all classify the site soil under seismic action in the relevant codes, and the corresponding codes are shown in Table 6-4.

Comparison of code system **Table 6-4**

Zone	Code	The latest version		Applicable area
European code	Eurocode 8	Eurocodes PrEn 1998		Europe, after 2010
American code	Standard	ASCE7	ASCE7-2010	the Unite States
	Code	NBC	The BOCA National Building Code	Northeast of the Unite States
		SBC	Standard Building Code	Southeast of the Unite States
		UBC	UBC-1997	West of the Unite States
		IBC	IBC-2009	the Unite States
Chinese code	Code for seismic design of buildings		GB 50011—2010	China

According to the different properties of soil layers, Eurocode 8 (EN 1998-1) divides the site under earthquake action into five basic types: A, B, C, D and E, as shown in Table 6-5. Two special types S_1 and S_2 are defined, namely highly plastic soil and liquefiable soil. According to the European code, site classification should first be carried out according to the $v_{S,30}$ value in Table 6-5 (average shear wave velocity of soil within 30m of the covering layer (shear strain is less than 10^{-5})). If the $v_{S,30}$ value cannot be provided, it should be carried out according to the N_{SPT} value (hammer number of standard penetration test). For S_2 type sites, the possibility of soil destruction under seismic action should be considered.

Site classification in Eurocode 8 **Table 6-5**

Site classification	Soil layer description	$v_{S,30}$	N_{SPT}	C_u
A	Rock or other similar rock formations, including weak overburden up to 5m thick	>800	—	—
B	Very dense sand, gravel, or hard soil are at least a few tens of meters thick and its mechanical properties gradually increase with the depth	360~800	>50	>250
C	Deep compacted or moderately compacted sand, gravel, or hard clay ranging in thickness from tens to hundreds of meters	180~360	15~50	70~250
D	Loose to medium-compacted non cohesive soils (with or without soft clay layers) or soft to hard clay	<180	<15	<70
E	It contains the soil layer composed of hard materials with a shear wave velocity of C or D type soil, a thickness of 5~20m, and a shear wave velocity greater than 800m/s	—	—	—
S_1	A layer of soft clay or silt at least 10m thick, having a high ductility index (PI>40) and high-water content	<100	—	10~20
S_2	Liquefiable soil layer, sensitive clay layer, or any other soil layer not included in A~E or S_1 type	—	—	—

Note: C_u is the undrained shear strength of the soil.

As one of the commonly used codes in the United States, UBC is short for "Uniform Building Code" and published by BOCA, one of the three publishing organizations of structural codes in the United States. Both UBC-1997 and IBC are considered according to a relatively fixed covering thickness of 30.48m (100ft). According to the shear wave velocity of the soil layer, the site is divided into 5 categories, as shown in Table 6-6. The difference is that the shear wave velocity of the soil layer is slightly different.

Site classification by the United States code UBC and IBC **Table 6-6**

Soil type	Soil type	Shear wave velocity of soil layer v_s(m/s)	
		UBC	IBC
Hard rock	SA	>1500	>1524
Rock	SB	760~1500	762~1524
Soft rock and dense soil	SC	360~760	365.8~762
Hard soil	SD	180~360	182.9~365.8
Soft soil	SE	≤180	≤182.9

According to China's code for seismic design of buildings GB 50011—2010, construction sites are divided into four categories according to the equivalent shear wave velocity of soil layer and the thickness of site overburden, among which the thickness of site overburden can be changed. Type I is classified into two kinds of subcategories: Type I_0 and Type I_1. When there is a reliable shear wave velocity and the thickness of the overburden and its value is near the boundary of the site category listed in Table 6-7, the interpolation method should be allowed to determine the design characteristic period used for seismic action calculation.

Overlay thickness of different site type(m, GB 50011—2010) **Table 6-7**

Shear wave velocity of rock or equivalent shear wave velocity of soil(m/s)	Site type				
	I_0	II_1	Ⅱ	Ⅲ	Ⅳ
$\nu_s > 800$	0				
$500 < \nu_s \leqslant 800$		0			
$250 < \nu_s \leqslant 500$		<5	≥5		
$150 < \nu_s \leqslant 250$		<3	3~50	>50	
$\nu_s \leqslant 150$		<3	3~15	15~80	>80

Note: ν_s in the table is the shear wave velocity of rock.

It can be seen from the comparison that among the three classifications, China's code for seismic design of buildings GB 50011—2010 has the least classification of soil. The main difference between the Chinese code and the European code is reflected in the different methods adopted for the division of site types. There are seven site types in Eurocode 8, which are defined in particular for special types such as highly plastic clay, liquefiable soil and sensitive clay. Compared with the Chinese seismic code, the division of site types is more detailed and clearer.

The classification of site types in China's code for seismic design of buildings GB 50011—2010 is basically the same as that in the United States code UBC-1997. The main difference is reflected in the specific value of shear wave velocity of soil layer used for geotechnical classification, resulting in the local crossover phenomenon of the classification of site soil types by the two codes.

6.4 Comparison of seismic acceleration and regional difference at home and abroad

Chinese code GB 50011—2010 divides earthquake intensity into 12 degrees.

American code UBC1997 listed by the designed basic seismic acceleration values are divided into zones 1, 2A, 2B, 3, and 4. Table 6-8 shows the comparison between the Chinese code GB 50011—2020 and the American Code UBC-1997. According to the corresponding relationship between seismic fortification intensity and the designed basic seismic acceleration value given in article 3.2.2 of Chinese code GB 50011—2010, there is no corresponding partition for seismic fortification intensity 6 in the United States code UBC-1997, and the seismic fortification intensity 7 and 8 in Chinese code GB 50011—2010 correspond to two different acceleration values in the United States code UBC-1997, as shown in Table 6-8.

The corresponding relationship in seismic intensity, zoning and basic acceleration value

Table 6-8

The basic seismic acceleration value	0.05g	0.075g	0.10g	0.15g	0.20g	0.30g	0.40g
Chinese code GB 50011—2010	6 degree	6.5 degree	7 degree	7 degree	8 degree	8 degree	9 degree
American code UBC-1997	—	1	—	2A	2B	3	4

Thel type in American Code UBC-1997 is not corresponding to the 6 degrees in Chinese code GB 50011—2010. In the comparison, the linear interpolation of the accelerations in the 6 degree region (0.05g) and the 7 degree region (0.10g) was performed artificially, and the obtained basic acceleration value of 0.075g was defined as the 6.5 degree region, as shown in Table 6-8.

The American code UBC-1997 lists the seismic zoning of major Chinese cities, as shown in Table 6-9. It can be seen from the table that the corresponding relations listed in Table 6-9 are different from those in Table 6-8, and the change trend of strength is consistent.

Zoning comparsion between America code UBC-1997 and Chinese code GB 50011—2010

Table 6-9

Code	Beijing	Chengdu	Guangzhou	Nanjing	Qingdao	Shanghai	Shenyang	Taiwan
American code UBC-1997	4	3	2A	2A	3	2A	4	4
Chinese code GB 50011—2010	8 degree 0.20g I group	7 degree 0.10g I group	7 degree 0.10g I group	7 degree 0.10g I group	6 degree 0.05g II group	7 degree 0.10g I group	7 degree 0.10g I group	8～9 degree 0.20g～0.40g I&II group

For a given Chinese city, its basic seismic intensity and seismic grouping are determined according to China seismic parameter zoning map (GB 18306—2015) and code for seismic design of buildings (GB 50011—2010), so it has a corresponding relationship with the ground motion parameters in American code IBC-2003. The spectral acceleration values of Chinese seismic zones in American codes can be obtained by calculation (Table 6-10 and Table 6-11).

Spectral acceleration values of Chinese seismic zones in United States code $Ⅱ_c$ (g) **Table 6-10**

Parameter name	7 degree 0.10g	7 degree 0.15g	8 degree 0.20g	8 degree 0.30g	9 degree
S_s	0.65	1.01	1.13	1.38	1.58
S_1	0.16	0.22	0.25	0.33	0.39
S_2	0.18	0.26	0.30	0.39	0.48
S_3	0.21	0.30	0.35	0.46	0.55

Note: 1. $Ⅱ_c$ refers to the part of class II sites regulated by Chinese code that is equivalent to class C sites regulated by the American code.
2. S_s and S_1 are spectral accelerations of class B ground in short period and 1s period under the maximum earthquake consideration in IBC-2003.

Spectral acceleration values of Chinese seismic zones in United States code $Ⅱ_D$ (g) **Table 6-11**

Parameter name	7 degree 0.10g	7 degree 0.15g	8 degree 0.20g	8 degree 0.30g	9 degree
S_s	055	0.89	1.04	1.38	1.58
S_1	0.11	0.17	0.20	0.26	0.31
S_2	0.13	0.20	0.23	0.31	0.40
S_3	0.15	0.24	0.28	0.38	0.46

Note: $Ⅱ_D$ refers to the part of class Ⅱ sites regulated by Chinese code that is equivalent to class D sites regulated by the American code.

In the ground motion parameter zoning, the probability level adopted by the Chinese code GB 50011—2010 and the European code PrEN1998-1 is the same, both exceeding the probability by 10% within 50 years, and the corresponding recurrence period is 475 years. Therefore, the design ground motion parameters of the two codes can be generally used without the problem of conversion of recurrence period. However, due to the big difference between the two codes in the standard site used in seismic zoning, the site effect should be paid attention to in the use process.

Ground motion parameters in a given Chinese city in the European PrEN1998-1 specification can be determined by the following equation:

$$a_g = A_{cc} / S$$

a_g——The peak ground motion acceleration of the earthquake designed on the A-class ground in PrEN 1998-1 is shown in Table 6-12;

S——Site coefficient of B class or C class in PrEN 1998-1 corresponding to Ⅱ sites in Chinese code GB 50011—2010;

A_{cc}——Basic peak ground motion acceleration corresponding to basic intensity in Chinese code GB 50011—2010.

The a_g value of the Chinese zoning in PrEN-1998 **Table 6-12**

Magnitude	Site type	7 degree 0.10g	7 degree 0.15g	8 degree 0.20g	8 degree 0.30g	9 degree
M>5.5	Ⅱ$_B$	0.083	0.125	0167	0.250	0.333
	Ⅱ$_C$	0.087	0.130	0.174	0.261	0.348
M≤5.5	Ⅱ$_B$	0.074	0.111	0.148	0.222	0.296
	Ⅱ$_C$	0.067	0.100	0.133	0.200	0.267

Note: Ⅱ$_B$、Ⅱ$_C$ are class Ⅱ sites in Chinese code and class B and C sites in the European code respectively.

6.5 Thinking questions

Do you know of any other differences among Chinese codes, European codes and American codes?

Chapter 7 Appreciation of the world geological miracle

7.1 Great Ocean Road and the Twelve Apostles in Australia

7.1.1 Great Ocean Road

Great Ocean Road is located in western Melbourne, Australia. It was built to commemorate the soldiers who participated in the First World War (Figure 7-1). It started in 1919 and was completed in 1932.

Figure 7-1 Great Ocean Road Entrance Monument (Shot in November 2016)

The Great Ocean Road starts at Torquay and travels 276 kilometres westward to finish at Allansford, from Geelong, Lorne, Apollo Bay, Port Campbell to Warrnambool. The Great Ocean Road stretches along the west coast of Victoria (Figure 7-2). It clearly divides the highlands into two parts: the south and the north. The southern seaside part of Great Ocean Road still maintains its original ecology and full of shrubs. The northern side is a green area with a large area of pastures and woodland.

Figure 7-2 The Great Ocean Road (Shot in November 2016)

The spectacular scenery along the Great Ocean Road is endless (Figure 7-3), this road also is praised as one of the ten most beautiful roads in the world. The most beautiful view of the Great Ocean Road is the sunset with birds fly, the golden sun shines on the straight cliffs, on the endless blue sea, wave constantly comes, hitting the rock made a crisp sound and splattered white foam (Figure 7-4).

Figure 7-3 Broken London Bridge in Great Ocean Road (Shot in November 2016)

Figure 7-4 Sunset over the cliff of Great Ocean Road (Shot in November 2016)

7.1.2 The Twelve Apostles

The Twelve Apostles is located on the shoreline of Port Campbell National Park, about 220 kilometers from southwest of Melbourne, Australia.

Over the past 10 million to 12 million years, limestone has been eroded by storms and sea breeze from the South Pacific and weathering has continued. Many caves have formed on limestone cliffs. Over time, limestone caves are arched from small to large, and until collapsed. These huge rocks, about 50m in height, were separated from the shore and formed the scene of the "the Twelve Apostles". The twelve broken rocks standing in the azure ocean are shaped like human faces and are known as the Twelve Apostles. It is a famous landmark on Australia's Great Ocean Road.

As time passed, five of the Twelve Disciples have collapsed in the past few decades. The most recent collapse occurred on July 3, 2005, when a rock of up to 45 meters suddenly collapsed and became gravel within seconds. The fallen gravel is still 10 meters higher than the sea level. Someone has predicted that the Twelve Apostles will all disappear after several years, only memories will be left.

7. 2 Yellowstone National Park, USA

Yellowstone Park is short for Yellowstone National Park and is under the responsibility of the National Park Service. On March 1, 1872, it was officially named the National Park for the protection of wildlife and natural resources. It was added into the World Natural Heritage List in 1978. Yellowstone Park is the core area of the "Big Yellowstone Ecosystem", while the "Big Yellowstone Ecosystem" is the most complete and largest temperate ecosystem on Earth.

Yellowstone National Park covers an area of about 9000km^2, mainly located in Wyoming, USA and partly in Montana and Idaho. Most of terrains are open igneous rock plateaus. The Yellowstone volcano caldera is located in the red line area in Figure 7-5. It is centered on the West Thumb on the west side of Yellowstone Lake, 15 miles to east and to west, and 50 miles to south and to north, forming a huge volcanic crater. A magma pool with a diameter of about 70 kilometers and a thickness of about 10 kilometers is under the crater. The huge magma pool is only 8 kilometers away from the ground and continues to expand.

Figure 7-5 Distribution map of Yellowstone Park
(Source from http://www. yellowstoneparknet. com/)

7. 2. 1 Yellowstone supervolcanos group

Supervolcanoes haven't an official definition up to now; it was popularized by a documentary film photographed by the BBC in 2000. Some scientists use this name to describe volcanoes that are super strong and have a particularly large eruption. Yellowstone supervolcanos group had erupted on a large scale about 2. 1 million years ago, 1. 3 million years ago and 640000 years ago, three large-scale eruptions formed three famous tuff sites in the Yellowstone National Park. The latest time of large-scale eruption was 638000 BC.

The Yellowstone caldera was formed by the eruption of 640000 years ago. The eruption released 1000 km^3 of volcanic ash, rocks and pyroclastic materials, forming a crater about 1000m deep and 84000m×45000m in plan size, and depositing a Lava Creek Tuff.

The most violent eruption of Yellowstone volcano occurred about 2. 1 million years ago. It emitted 2450 km^3 of volcanic material and produced rocks called the Huckleberry Ridge Tuff, which formed the Island Park Caldera.

The Yellowstone volcano emitted 280 km^3 of volcanic material 1. 2 million years ago, forming Henry's fork caldera and depositing Mesa Falls tuff.

There are thousands of small earthquakes in Yellowstone Park each year. Since history has been recorded, there have been six earthquakes of magnitude 6 or above, including the magnitude 7. 5 earthquake that occurred outside the northwest boundary of Yellowstone Park in 1959. The earthquake triggered a huge landslide that caused part of the dams in Lake Hebgen to collapse. The landslides blocked the river downstream and formed a new lake, the Earthquake Lake. On June 30, 1975, a magnitude 6. 1 earthquake occurred in Yellowstone Park. In 1985, scientists monitored more than 3000 relatively minor earthquakes in the northwest of Yellowstone Park for three months. In the 7 days since April 30, 2007, scientists have detected 16 small earthquakes of magnitude 2. 7 at the Yellowstone caldera. In December 2008, during the four-day period, scientists monitored more than 250 earthquakes under Yellowstone Lake, the largest of which was 3. 9. In January 2010, scientists monitored more than 250 earthquakes within a two-day period. From June 12 to mid-August 2017, scientists have monitored more than 1500 earthquakes, and the magnitude of the largest earthquake was 4. 4. Seismic activity in Yellowstone National Park is continuing.

7. 2. 2 The Grand Prismatic Spring

The Grand Prismatic Spring is an excellent representative of Yellowstone Hot Springs. It is also the largest hot spring in the United States and the third largest in the world. It is known as the most beautiful plane. The Grand Prismatic Spring is about 75m to 91m wide and 49m deep. There are about 2000 liters of underground water flowimg out per minute, which has a maximum temperature of 85℃.

The Grand Prismatic Spring display blue, green, yellow, orange and red colors from inside to outside (Figure 7-6). Due to the water temperature is different, the bacteria with different colors live, so the color also changed as a concentric circle. The beauty of the Grand Prismatic Springs is that the color of the lake changes with the change of seasons. In summer, it appears orange, red or yellow. In winter, the water is dark green. The Grand Prismatic Springs under the sunset are even more beautiful.

Figure 7-6 The Grand Prismatic Spring (Shot in July 2018)

7. 2. 3 Yellowstone National Park Grand Canyon

The Yellowstone River runs through volcanic rocks. Due to the long-term erosion and transportation of rivers, the magnificent Yellowstone Grand Canyon is formed. The canyon is located between the fishing bridge and the tower, has 60m deep, 200m wide, and about 32km long. It is extremely dangerous (Figure 7-7).

Figure 7-7 The Yellowstone National Park Grand Canyon (Shot in July 2018)

The most fascinating of the Yellowstone River is neither the depth nor shape of the canyon, nor the torrential waterfalls, but the eroded weathered volcanic rocks. The entire canyon walls glow with dazzling luster, especially in the sun. It is dazzling, showing white, yellow, green, blue, vermilion and other colors. Millions of tons of rocks seem to have been painted by oil paint.

Between 8 : 00 and 10 : 00 in the morning, overlooking the Grand Canyon from the Artist Point, the sunshine crosses the southern peaks and simks on the canyon. The Grand Canyon is particularly powerful under the shadow and light.

The famous sights in Yellowstone Park include the magnificent Old Faithful Geyser (Figure 7-8), the peaceful Yellowstone Lake and so on.

Figure 7-8 Old Faithful Geyser (Shot in July 2018)

7.3 Indonesia Volcano Merapi

Merapi means "endless fire". Merapi is a conical volcano located on the volcanic peak near the center of Java, Indonesia, 32 kilometers northeast of Yogyakarta, with an elevation of 2911 meters. Volcano Merapi is one of the 130 most active volcanoes in Indonesia. Since 1548, it has erupted 68 times; it is also the youngest of the volcanoes in south Java.

7.3.1 The historical record of eruptions of Volcano Merapi

Through stratigraphic analysis, it was proved that the Merapi volcano had erupted

400000 years ago. Until 10000 years ago, it has been typical lava jets to form basalt. An explosive eruption was later formed in this area. The magma was a more viscous andesite and formed a lava dome. The collapse of the lava mound would cause a pyroclastic stream and a more violent eruption to form an eruption column.

The Merapi Volcano usually has a small eruption every 2 to 3 years and a large eruption every 10 to 15 years. The larger eruptions occurred in 1006, 1786, 1822, 1872, 1930 and 1976.

The Merapi Volcano erupted for the first time in 1006, crushing the top of the mountain, destroying a kingdom in the middle-jaw island and burying the Borobudur Temple Compunds, one of the Seven Wonders of the World (Figure 7-10). Until the early 19th century, the Borobudur Temple Compunds was cleared out. It was also known as the Four Wonders of the Ancient East with the Great Wall of China, the Taj Mahal of India and the Angkor Wat of Cambodia.

Figure 7-9 Repaired Borobudur Temple Compunds (Shot in August 2016)

The eruption of Merapi Volcano in 1872 was the largest, and it caused the destruction of 13 villages and 1400 deaths with the eruption in 1930.

The eruption of Merapi Volcano in 1992 lasted for 10 years. The lava mound continued to rise. By 1994, the lava dome reached the height of crater. The collapse of the entire lava dome formed several thousand kilometers of gravel steam, resulting in 43 people deaths.

On May 30, 2006, Merapi Volcano repeatedly emitted smoke and lava. On November 5, 2010, Merapi Volcano erupted again. The diffuse volcanic ash caused visibility in some parts of Yogyakarta less than 20 meters.

Merapi Volcano can be seen from the Kali Adem viewing platform by jeep. The viewing platform is about five kilometers from the top of the volcano. Along the route is volcanic debris (Figure 7-10). Houses, household items and cattle skeletons which left after the 2010 Merapi Volcano eruption are displayed in the Mini Museum. A clock is fixed at the moment of volcanic eruption (Figure 7-11).

Figure 7-10 Gravel along the Merapi Volcano (Shot in August 2016)

Figure 7-11 Broken clock after the eruption of Merapi Volcano on 2010 (Shot in August 2016)

7.3.2 Volcanic monitoring

Due to the Merapi Volcano has always threatened the safety of human settlements; it is listed by the International Association of Geochemistry and Volcanology as one of the 16 global volcanoes that should strengthen monitoring and research.

The monitoring of seismic waves of Merapi Volcano began in 1924. And it was confirmed that a strong seismic activity would happen before the volcanic eruption through the 1930 eruption.

At present, a monitoring network consisting of 8 seismographs is set up around Merapi Volcano. Based on the monitoring data, the hypocenter can be determined. The hypocenter that is below 1.5 km from underground will not be detected. In addition to the monitoring of earthquakes, geomagnetic monitoring is also performed, and changes in geomagnetism also reflect the fluctuation of magma, which has a positive effect on the prediction of volcanic eruptions.

7.4 Giant Crystal Cave

Crystal Cave is a space cave in soil or rock. Due to the water infiltrating through the gap, together with the combined effect of temperature, pressure and time, the cave begins to grow crystals and eventually forms a crystal cave.

7.4.1 The discovery of Giant Crystal Cave

The meaning of Naica in Tarahumara is a dark place. The Naica mine was first discovered by miners in Chihuahua in 1794. They discovered a vein of silver at the bottom of the Naica Mountains in the Chihuahua Desert. Until 1900, this deposit mainly produced gold and silver, and it was only in 1900 that it began to explore a large scale of zinc and lead.

In 1910, the miners discovered another special cave, the Sword Cave, under the Naica Mountains. The wall of cave was filled with huge sword-shaped crystals. The sword cave is hidden 120m underground; the crystals are about 1m long.

In 2000, two miners who worked for Peoles Industrial discovered the Naica Crystal Cave while digging an underground tunnel. The cave contains many of the largest natural crystals in the world. These translucent crystals are 11 meters long and weigh 55 tons. This crystal cave is called the "Naica Crystal Cave," also known as the "Giant Crystal Cave," buried in 274m below the Naica Mountains in the Chihuahua Desert. The Naica Crystal Cave is a horseshoe-shaped limestone cave with a width of 10 meters and a length of 30 meters. It is also the largest underground crystal cave in the world.

In December 2009, three scientists, Penelope Boston, Michael Spielder and Danielle Winget collected samples from a water hole in the Naica Crystal Cave and performed indoor experiments. The results show that the water in this crystal cave contains a lot of virus.

7.4.2 The cause of formation of Naica Crystal Cave

26 million years ago, the volcanoes in the Naica Mountains were very active. They were filled with high-temperature anhydrous gypsum, which is a type of gypsum with low moisture content. When the magma under the mountains begins to cool down and the temperature drops below 58℃, the anhydrite gypsum is no longer stable and gradually decomposes. The sulfate and calcium ions in the water gradually increase, depositing in this cave for millions of years, and finally forming a giant transparent gypsum crystal.

The Spanish geologist Ruertz, who has been studying the cave, said: "Naica is very unique. The probability of having another Naica Crystal Cave on Earth is very small. We should protect this unique natural heritage."

In the Naica Crystal Cave, the temperature is up to 50℃ and the humidity is up to 100%, which are extremely deadly for humans. Special cooling suit equipped special respiratory systems must be worn before enter the cave. These difficulties are still unable to stop scientists from exploring the cave. The mystery of the Naica Crystal Cave is gradually unfolded.

7.5 The Giant's Causeway

The Road of Giant, also known as Giant's Causeway, is located on the Atlantic coast, about 80 km northwest of Belfast in Northern Ireland. It is a multi-kilometer causeway and was listed as a World Natural Heritage Site in 1986. It was also listed as the fourth natural wonder of the United Kingdom in 2005.

The Giant's Causeway is a product of eruption of ancient volcano. It consists of 37000 columnar basalts and is a perfect representation of the columnar basalt landscape. The cross section of the pillars that make up the Giants' Causeway is mainly hexagonal, with a width of 37～51cm, and a typical width of about 45cm. The coast is about 8000m in length. Some of the stone pillars are 6 m above the sea surface, and the maximal height can reach about 12 m. The thickness of solidified lava above these stone pillars is about 28 cm. There are also some stone pillars that are submerged underwater or at sea level.

The columnar basalt landscapes similar to the Giants' Causeway include the islands of Staffa in the Inner Hebrides, the south of Iceland, the Pillar Hill in Liuhe County, Jiangsu Province, and the Niutoushan Scenic Area in Zhangzhou, Fujian Province, China. But none is so complete and spectacular like the Giant's Causeway.

As the global warming and the rise of sea level, Giant's Causeway will suffer more violent waves and storm erosion. Some people predict that in the near future, it will be difficult to see the unique landscape on part of Giant's Causeway.

Chapter 8 Geology practice route

8. 1 Visit practice of China Geological Museum

8. 1. 1 Purpose of internship

(1) To understand the basic properties of major rock-forming minerals and rocks.

(2) To understand the occurrence mechanism and phenomenon of common geologic disasters.

(3) To understand the method of identification of gems.

8. 1. 2 Practice route and content

China Geological Museum is located in Beijing West Four Mutton Alley 15th, including four floors (Figure 8-1), the first floor is for the Earth Hall, the second floor is for the Mineral and Rocks Hall, Gem and Jade Hall, the third floor is for the Prehistoric Creatures Hall, the fourth floor is for the Land and Resources Exhibition Hall. The Tour route can start at the first floor, up to four floors, or from any level.

The main contents of the internship include:

(1) To understand the physical properties of the main rock-forming minerals on the first floor of the Earth's hall.

(2) To understand the nature of the main rocks in the Minerals and Rocks Hall and learn about gemstone characteristics and simple identification methods in Gem and Jade Hall on the second floor.

(3) To understand the occurrence mechanism and phenomena of geological disasters such as earthquake, glaciers, rivers, karst and other geological hazards on the second and third floor.

(4) To understand the national leaders and geological experts contributions and related geological work in the Land and Resources hall on the fourth floor.

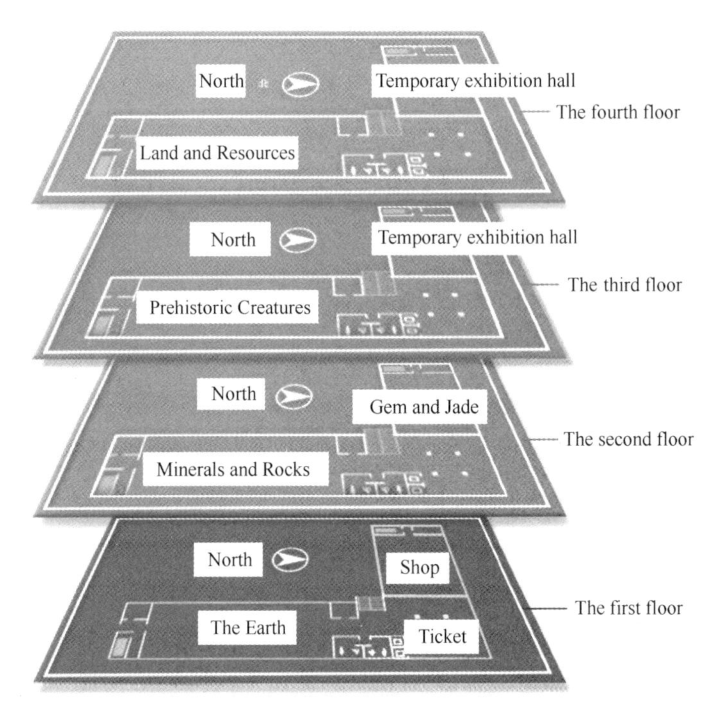

Figure 8-1 China Geological Museum layout diagram (Source from http://www.gmc.org.cn/)

8.2 Basic geology training practice in Green Garden, Beihang University

8.2.1 Purpose of internship

(1) Familiar with the use of geological compass instrument.

(2) To understand the measurement method of slope angle.

8.2.2 Internship route and content

The Green Garden of Beihang University is located on the west side of the library of the College Road Campus, with small rock slopes and soil slopes, which can meet the learning requirements of the geological compass and the measurement requirements of the slope angle. Internship route from the southeast corner of Green Garden, along the Lotus Pond westward about 30m to reach the Arch Bridge, the west side of the Arch Bridge 1th, after the arch bridge forward about 20m to reach the second hillside, for No. 2 slope. The basic training of geological practice around two mountain slopes, when encountering special circumstances, can advance to the other slopes of Green Garden for basic training, including observing the lithology of the slope, measuring the occurrence of the specified strata and measuring the slope angle, and finally returning to the southeast corner of the Green Garden(Figure 8-2,Figure 8-3).

Figure 8-2 Green Garden Arch Bridge (shot in August 2017)

Figure 8-3 Green Park, No. 1 slope in Green Garden (shot in August 2017)

The main internship content includes:

(1) The use of geological compass instrument training, master the magnetic declination correction of geological compass, measure any slope occurrence.

(2) Measure the slope angles of No. 1 and No. 2 Slope, consider and master the method of measuring irregular slope angle.

(3) Record the practice content and master the record format of the observation con-

tents of geological points.

8.3 The geology practice route from Yexi to Xiehejian

8.3.1 Purpose of internship

(1) To observe the fold, including box fold and flexible fold.

(2) To observe the thick limestone in the Majiagou Group of Ordovician system (O2).

8.3.2 Internship route and observation content

Yexi Bridgehead-Datai route, Fengsha route and Yexi-Xiehejian route belong to a Miaofengshan town in Mentougou District, Beijing, located in the northeast of Mentougou District, only 30 kilometers east of Beijing city center. Yongding River and China National Highway 109 cross this town. Xiehejian is located on the hill near the Yongding River. The Fengsha railway route runs between the town and the Yongding River. There are 11 observation points on this route, which are detailed as follows:

(1) The 1th observation point is Xiehejian. Observe the geomorphologic features of the two sides of the Yongding River.

Under the action of river erosion, the Yongding River has different riparian characteristics. The concave bank and the convex bank alternate with each other. The river valley terrace developed well and formed the obvious secondary river terrace.

(2) The 2nd observation point: Observe the features of box fold on the right bank following the stream of the river.

Miaofengshan locates at the right bank of the river and is composed by three synclines and two anticlines. The box fold and one side of the box core are easy to see on the right bank.

(3) The 3rd observation point: Cross the right tunnel of the Datai route in the wings of the syncline and observe its surrounding rock.

The tunnel entrance locates at the core of a syncline, whose shape like the horseshoe and has not lining. The surrounding rock of tunnel is a thick layer of limestone in the Ordovician system (O2).

(4) The 4th observation Point: Observe the differences between these two exits of tunnels.

Another exit of the 1th tunnel has regular shape and is supported with lining.

(5) The 5th observation point: Observe the rock mass along opposite direction of the river flow.

There is fault structure in this segment, because shear moving takes place in a thin limestone layer of the thick rock mass. The thin layer of limestone is suddenly faulted along the strike direction. The dikes are cut off and there are tectonic rocks.

(6) The 6th observation point: Observe the contact relationship between magmatic rocks and sedimentary rocks.

The contact relationship of sedimentary rocks and magmatic rocks should be judged according to the contact surface characteristics of the rock wall with a width of about 50cm.

(7) The 7th Observation Point: Observe the artificial shoulder wall.

(8) The 8th Observation Point: Observe the artificial spur dike on sides of the Yongding River.

(9) The 9th Observation Point: Observe the accumulation characteristics of the Yongding River near the spur dike.

(10) The 10th observation point: Observe the mylonite, fault striations and phyllite out of tunnel in the fracture zone that locates at the entrance of the 2nd tunnel.

(11) The 11th Observation Point: Observe the full lining support of the whole 1st tunnel body.

The 1st tunnel adopted full lining support and laid drainage ditches on both sides of the entrance, because it crosses the core of Jiulongshan syncline and leakage seriously.

8.4 The geology practice route from Nankou to Qinglong Bridge

8.4.1 Purpose of internship

(1) To understand the identification method of rock weathering degree.

(2) Observe faults and joints, and classify them.

(3) Observe the masterpiece of Zhan Tianyou.

8.4.2 Internship route and observation content

This route is from the Guangou of Nankou Town in the south to the Badaling Great Wall in the north, with total of 8 observation points, as follows:

(1) The 1st observation point: observe the normal fault.

The Guangou is the starting point of observation, the 1st observation point, where the normal fault can be observed.

(2) The 2nd observation point: Observe the Sinian sedimentary rocks.

To the west side of the ditch, the small cave on the west side of the gully is the 2nd observation point, the Sinian sedimentary rock (siliceous limestone) and the pink syenite dyke.

(3) The 3rd observation point: Observe the small rock mass at the foot of the moun-

tain slope.

The way forward, from the 2nd observation point about 500m there is a fork junction, along the forkslope; you can observe the steep-dip rock mass. There is the 3rd observation point. There is a small size of the rock mass at the foot of mountain slope, which is Sinian Metamorphic Rocks (quartzite), according to observing the color and weathering degree of rock mass.

(4)The 4th observation point: Observe pattern joints and complete the joint investigation.

Return the artificial path; go down the ditch about 50m ahead, then go upstream along the valley, along with springs exposed, get the 4th observation point. Bulk quartzite rock rolling down from the mountain and the exposed bedrock with pattern joints can be seen here. The shear joints are obvious and have straight flat, smooth and dense shear plane, which is related to the shear activity of the slow shear force under the action of natural high ground stress.

It is necessary to complete the field investigation on joint occurrence, density, thickness, extension and joint width.

(5) The 5th observation point: Observe magmatic rocks.

Move on, the dip angle of the rock stratum becomes gradually gentle and the formation lithology changes; changes cover interface in vegetation is not obvious, there is the 5th observation point. The rock here is Archean gneiss, dark gray and has obvious gneissic structure. Its main mineral is amphibolites with a small amount of plagioclase and quartz.

(6) The 6th observation point: Observe the basement terrace, which is the typical Binary structure characteristics of fluvial deposit.

Continue along the ditch, the exposed first terrace is the 6th observation point. The bare bedrock is visible at the bottom, which is covered with the gravel layer and the soil layer. This is a typical binary structure characteristic in the river sediments.

Gravellayer and bedrock is steep, which indicates that the strong crustal movement made the valley bottom up rapidly and river cut downwards strongly.

(7)The 7th observation point: Observe the artificial diversion dike.

Enter the highway bridge along the valley, artificial diversion dike locates at the outside of the bridge, where is the 7th observation point.

The direction of the embankment is generally parallel to the bridge embankment. The purpose is to reduce the scour to the bridge embankment during the flood period, so as to divide the water flow.

(8)The 8th observation point: Observe magmatic rock.

Drive near to the Badaling martyrs cemetery, then walk to the QingLong Bridge station, where is the 8th observation points. Yanshanian granite porphyry can be observed on the way, which has fleshy red fresh surface, massive structure and and holomorphic granular structure, some of them are porphyry structures. Its main mineral components are orthoclase, quartz and a small amount of amphibolite and black mica. It is formed due to the condensation of magma under the ground.

The rock masshere is weathered and denudated strongly, whose granite surface is yellowish-brown and the luster is dim. The weathered rock is loose sand granular and has decreased strength. The vegetation is directly grown in the crushed granite, and the biological weathering effect is remarkable.

The middle ground space at QingLong Bridge station is extremely limited. You can see the famous railway crossing which is still operating normally.

8.5 Geology practice route from Gubeikou Beijing Tongliao Railway Chaohe crossing to Taoshan tunnel

8.5.1 Purpose of internship

(1) Observe and identify the three kinds of rocks.

(2) Observe and classify faults and joints correctly.

(3) Observe geological process of river.

(4) Understand landslides, collapse and so on.

8.5.2 Internship route and observation content

Internship route locates at the junction of Gubeikou town and Bakeshi town Luanping County, facing with the famous Juyongguan Pass. Panlong mountain locates at the east part of the north Gubeicheng town, Wohu mountain locates at the weat part, all of them are precipitous and steep. Chao River flows into Miyun Reservoir from the northern valley fjord by Gubeikou.

There are faults and joints, three major rock types, river geological process, landslides, collapse and other engineering geological disaster. The exposed stratums here mainly include sandstone, conglomerate, and shale of Jurassic system, dolomite limestone and quartzite of Sinian quartzites and gneiss of Archean system. This internship route with a total of five observation points, as detailed below:

(1)The 1st observation point: Observation Wohu Mountain and Gubeikou basin.

The 1st observation point is the Gubeikou train station. Wohu Mountain is mainly composed by Archean metamorphic rocks, Mesoproterozoic stratum, late Proterozoic stratum, Mesozoic strata, Mesoproterozoic intrusive rocks and late Proterozoic intrusive rocks. The contact relationship of intrusive rocks here needs to be observed.

The boundary of Gubeikou basin is mainly controlled by the Gubeikou fault. Differences of the thick Tucheng subgroup conglomerate, sandstone and mudstone deposited in this basin need to be observed.

(2)The 2nd observation point:Observe the lithology of Wohu Mountain.

There are Jurassic sandstone, conglomerate, shale, Sinian dolomitic limestone and quartzite, and Archean gneiss.

(3)The 3th observation point:Observe the Wohu mountain fault.

The strike direction of Wohu Mountain fault is from northeast to southwest, dip direction is northwest, and dip angle is close to a right angle. The upper wall is Sinian dolomitic limestone and footwall is Jurassic conglomerate. It is a reverse fault judging from relative displacement between the two walls.

(4)The 4th observation point:The big bend of Chaohe, observe the erosion and deposition process of river.

The Chaohe flows from east to west, and washes the coastal slopes fiercely. The wash process is more intense during the flood season, which causes the mountain slope to be hollowed out and collapses.

There are several gullies in the mountain near the river side. The rock mass on both sides of the gully wall is badly weathered and broken so that collapse often occurs. The collapsed material was washed out of the ditch and formed a diluvial fan.

(5)The 5th observation point:Observe the landslides and collapses.

8.6 Geology practice route in Tianjin Jixian National Geological Park

8.6.1 Purpose of internship

(1) Observe the three rock types and geological structure.

(2) Observe the contact relationship of the sedimentary rock formation and features of bedding plane.

8.6.2 Internship route and observation content

Tianjin Jixian National Geological Park locates in Jixian County, Tianjin.

Middle Upper Proterozoic stratigraphic profile from Chanzhou village to Fujunshan, which is 24 km long, represents up to 1 billion years (18~800million years ago) geological history. It is a complete Marine and land transition process in geological history. The paleontological fossils are rich here. The Macroscopic multicellular fossil found here led the age of multicellular organisms bring forward from internationally recognized 900 million years to 1.7 billion years.

The strata of the Middle Upper Proterozoic stratigraphic profile in jixian are complete, expose continuously, and in good keep condition. Its geological structure is simple, the bottom line and top line are clear. It has shallow metamorphism.

Here is the oldest geological phenomenon in the world, such as the sedimentary sepiolite deposit and iron in 12～13 billion years ago, and the oldest vesicular structure in the world found in the Ridge group.

Jixian County in the late Proterozoic strata section is complete, exposed continuous, well preserved, simple structure, top and bottom boundaries clear and shallow metamorphic. Here are the most ancient geological phenomena in the world, such as the formation of sedimentary sepiolite deposits and iron from 12～13 billion years, and fumaroles of the oldest in the world found Ling group structure. This route has a total of 6 observation points, as follows:

(1)The 1st observation point: Observe the Archean (Arw) garnet gneiss and leucolite.

(2)The 2nd observation point: Observe the angular unconformity contact relationship, oblique bedding and cross bedding of the interface between Proterozoic and Archean.

The geological age of the Changzhou gutter group Changchengian System at the bottom of the Middle Proterozoic section in Jixian is about 1. 8 billion years. But the age of metamorphic rocks in the lower part of Archaean group is about 2. 5 billion years. There is an obvious gap in time, missing about 700 million years. The second observation point is a typical stratigraphic cut-off point, and the ancient weathering crust is visible here.

(3) the 3th observation point: Observe shale, angular unconformity, oblique bedding and cross bedding.

The third observation point locates near highway in the south of Qingshanling village. The rock of the Changzhou gully group Changchengian System (Chc) and the Chch group is mainly composed of dark shale and silty (cloud) shale. The different contact relationships between sedimentary strata layer can be observed here.

(4) The 4th observation point: Observe mud cracks and weathering of dolomite.

The fourth observation point locates at the east of the big Hongyu ditch.

The Chch group and Chc group is 326m thick, whose upper part is composed of massive sandy dolomite and dolomite sandstone, whose lower part is the cyclothem composed by the black gray dolomite, Sand-laden dolomite and Gray leaf dolomite.

(5) The 5th observation point: Observe the igneous rocks and geological structure.

The fifth observation point locates in the big Hongyu ditch ahead 300m, which belongs to the Great Wall group and Dahongyu group. It is mainly composed of sandstone, sandy dolomite and dolomite, with volcano rock. This observation point is mainly the igneous rocks, such as granite, granite porphyry, basalt and volcaniclastic rock. This observation point can also observe the structure of magmatic rocks, including massive structure, rhyolitic structure, vesicular structure and amygdaloidal structure.

(6) The 6th observation point: Observe parallel unconformity contact relationship.

The sixth observation point locates in big Hongyu ditch ahead about 1. 4km, here is the the Chd group and Gaoyuzhuang group (Chg), mainly in siliceous bands, stromatolite dolomite and shale containing manganese. Here we observed two formation parallel unconformity contact relations. The parallel unconformity contact relationship between the two groups can be observed here.

Appendix 1 Rock names and symbols

1.1 Plutonic intrusive rocks

NO.	Name	Symbol	NO.	Name	Symbol
1	Unclassified ultrabasic rock	Σ	19	Granite	γ
2	Dunite	φ	20	Monzonitic granite	ηγ
3	Olivinite	σ	21	Alaskite	γκ
4	Pyrozenite and amphibolite	ψ	22	Granodiorite	γδ
5	Pyrozenite	ψτ	23	Biotite granite	γβ
6	Amphibolite	ψo	24	Moyite	ξγ
7	Serpentinite	ψω	25	Monzonite	η
8	Unclassified basic rock	N	26	Quartz monzonite	ηo
9	Gabbro	ν	27	Plagioclonal granite group	Γo
10	Olivine-gabbro	σν	28	Plagioclonal granite	γo
11	Norite	νo	29	Unclassified alkaline rock	E
12	Anorthosite	νσ	30	Pulaskite	χ
13	Diorite	δ	31	Alkaline rock	χξ
14	Quartz diorite	δo	32	Nepheline syenite	ε
15	Black mica diorite	δβ	33	Syenite	ξ
16	Orthoclase diorite	ξδ	34	Grano syenite	γξ
17	Gabbro-diorite	νδ	35	Quartz syenite	ξo
18	Undivided granite	Γ			

1.2 Hypabyssal intrusive rocks

NO.	Name	Symbol	NO.	Name	Symbol
1	Picrite-porphyrite	ωμ	12	Monzonitic granite porphyry	ηγπ
2	Gabbro-pegmatite	νρ	13	Orthophyre	ξπ
3	Gabbroporphyrite	νμ	14	Pegmatite	ρ
4	Diabase, sillite	βμ	15	Granite pegmatite	γρ
5	Dioritic porphyrite	δμ	16	Lamprophyre	χ
6	Granite porphyry	γπ	17	Anorthosite -Lamprophyre	δχ
7	Microcrystalline rock	τ	18	Minette	ξχ
8	Granite-aplite	γτ	19	Iolite	εχ
9	Quartz porphyry	oπ	20	Kimberlite	χσ
10	Granodiorite-porphyry	γδπ	21	Carbonatite	χC
11	Ivernite	ηπ			

1.3 Eruptive rocks

NO.	Name	Symbol	NO.	Name	Symbol
1	Unclassified ultrabasic eruptive rock	Ω	18	Rhyolite porphyry	$\lambda\pi$
2	Picrite	ω	19	Baschtaunite	$\lambda\chi\tau$
3	Unclassified basic eruptive rock	B	20	Felsophyre, felsite	$\upsilon\pi$
4	Spilite	$\mu\beta$	21	Pumice	$\upsilon\lambda$
5	Basalt	β	22	Quartz albitophyresi	$\lambda\varphi$
6	Picrite basalt	$\omega\beta$	23	nevadite	$\pi\lambda$
7	Basaltic glass rock	$\upsilon\beta$	24	Glass rock and aphanite	υ
8	Andesitic basalt	$\alpha\beta$	25	Unclassified sodium porphyry	Φ
9	Unclassified intermediate eruptive rock	A	26	Dacite-albitophyre	$\zeta\Phi$
10	Andesite	α	27	Andesite-albitophyre	$\alpha\Phi$
11	Andesitic porphyrite	$\alpha\mu$	28	Unclassified alkali eruptive rock	θ
12	Dacite	ζ	29	Phonolite	ν
13	Dacite porphyry	$\zeta\pi$	30	Trachyandensite	$\tau\alpha$
14	Dacite-porphyrite	$\zeta\mu$	31	Trachyte	τ
15	Keratophyre	$\chi\tau$	32	Trachy porphyry	$\tau\pi$
16	Unclassified acid eruptive rock	Λ	33	Alkali trachyte	$\chi\tau$
17	Rhyolite	λ	34	Alkali basalt	$\chi\beta$

1.4 Other rock symbols

NO.	Name	Symbol	NO.	Name	Symbol
1	Breccia	br	13	Lava	lv
2	Conglomerate	cg	14	Agglomerate	a
3	Sandstone	ss	15	Volcanic breccia	vb
4	Debris sandstone	ds	16	Tuff	tf
5	Siltstone	st	17	Agglomerate lava	al
6	Shale	sh	18	Breccia lava	bl
7	Clay (mud) rock	cr	19	Tufflava	tl
8	Mudstone	ms	20	Lava agglomerate	la
9	Limestone	ls	21	Lava breccia	lb
10	Marlstone	ml	22	Lava tuff	lt
11	Dolomite	dol	23	Slate	sl
12	Siliceous rock	si	24	Phyllite	ph

continued

NO.	Name	Symbol	NO.	Name	Symbol
25	Schist	sch	49	Alluviation	al
26	Gneiss	gn	50	Landslide accumulation	del
27	Orthogneiss	og	51	Soil slip accumulation	slf
28	Paragneiss	pg	52	Glacio-lacustrine deposit	lgl
29	Granulite	gnt	53	Wind-blown sand	col-s
30	Leptite	gnl	54	Aeolian loess	col-ls
31	Migmatite	mi	55	Eluvial deposit -Qdl	cld
32	Metasandstone	mss	56	Qdl-Dirt deposits	dls
33	Metamorphic volcaniclastic rock	mv	57	Lacustrine deposit	l
34	Meta-andesite	mas	58	Swamp deposit	f
35	Hornstone	hs	59	Marine accumulation	m
36	Marble	mb	60	Aeolian accumulation	e,eol
37	Greisen	gs	61	Chemical deposit	ch
38	Skarn	ssk	62	Biogenic accumulation	o
39	Hybrid rock	hr	63	Glacial deposit	gl
40	Cataclasite	tr	64	Glacial fluvial deposit	fgl
41	Structural breccia	sb	65	Artificial accumulation	a
42	Mylonite	ml	66	soil	pd
43	Phyllonite	pm	67	Random packing under gravity	col,c
44	Gossan	gnd	68	Qdl- diluvium	dpl
45	Elurium	el	69	Qdl- alluvial deposit	ald
46	Qdl	dl ,d	70	Alluviation- diluvium	alp
47	Colluvium	c,co	71	Alluviation-lacustrine deposits	lal
48	Proluvium	pl,prl	72	Agnostogenic accumulation	pr

Appendix 2　Apparent dip conversion table

True dip (A)	The included angle between strike direction of rock and profile (B−C)																
	80°	75°	70°	65°	60°	55°	50°	45°	40°	35°	30°	25°	20°	15°	10°	5°	1°
10°	9°51′	9°40′	9°24′	9°5′	8°41′	8°13′	7°41′	7°6′	6°28′	5°46′	5°2′	4°15′	3°27′	2°37′	1°45′	0°53′	0°10′
15°	14°47′	14°31′	14°8′	13°39′	13°34′	12°28′	11°36′	10°4′	9°46′	8°44′	7°38′	6°28′	5°14′	3°33′	2°40′	1°20′	0°16′
20°	19°43′	19°23′	18°53′	18°15′	17°30′	16°36′	15°35′	14°25′	13°10′	11°48′	10°19′	8°45′	7°6′	5°23′	3°37′	1°49′	0°22′
25°	24°48′	24°15′	23°39′	22°55′	22°0′	20°54′	19°39′	18°15′	16°41′	14°58′	13°7′	11°9′	9°3′	6°53′	4°37′	2°20′	0°28′
30°	29°37′	29°9′	28°29′	27°37′	26°34′	25°13′	23°51′	22°12′	20°21′	18°19′	16°6′	13°43′	11°10′	8°30′	5°44′	2°53′	0°35′
35°	34°36′	34°4′	33°21′	32°24′	31°13′	29°50′	28°12′	26°20′	24°14′	21°53′	19°18′	16°29′	13°28′	10°16′	6°56′	3°30′	0°42′
40°	39°34′	39°2′	38°15′	37°15′	36°0′	34°30′	32°44′	30°41′	28°20′	25°42′	22°45′	19°31′	16°0′	12°15′	8°117′	4°11′	0°50′
45°	44°34′	44°1′	43°13′	42°11′	40°54′	39°19′	37°27′	35°16′	32°44′	29°50′	26°33′	22°55′	18°53′	14°30′	9°51′	4°59′	1°0′
50°	49°34′	49°1′	48°14′	47°12′	45°54′	44°17′	42°23′	40°7′	37°27′	34°21′	30°47′	26°44′	22°11′	17°9′	11°41′	5°56′	1°11′
55°	54°35′	54°4′	53°19′	52°18′	51°3′	49°29′	47°35′	45°17′	42°33′	39°20′	35°32′	31°7′	26°2′	20°17′	13°55′	7°6′	1°26′
60°	59°37′	59°8′	58°26′	57°30′	56°19′	54°49′	53°0′	50°46′	48°4′	44°47′	40°54′	36°14′	30°29′	24°8′	16°44′	8°35′	1°44′
65°	64°40′	64°14′	63°36′	62°46′	61°42′	60°21′	58°40′	56°36′	54°2′	50°53′	46°59′	42°11′	36°15′	29°2′	20°25′	10°35′	2°9′
70°	69°43′	69°43′	68°49′	68°7′	67°12′	66°8′	64°35′	62°46′	60°29′	57°36′	53°57′	49°16′	43°13′	35°25′	25°30′	13°28′	2°45′
75°	74°47′	74°47′	74°5′	73°32′	72°48′	71°53′	70°43′	69°14′	67°22′	64°58′	61°49′	57°37′	51°55′	44°1′	32°57′	18°1′	3°44′
80°	79°51′	79°51′	79°22′	78°59′	78°29′	77°51′	77°2′	76°0′	74°40′	73°15′	70°34′	67°21′	52°43′	55°44′	44°33′	26°18′	5°31′
85°	84°56′	84°56′	84°41′	84°29′	84°14′	83°54′	83°29′	82°57′	82°15′	81°20′	80°5′	78°19′	75°39′	71°20′	63°15′	44°54′	11°17′
89°	88°59′	88°58′	88°56′	88°54′	88°51′	88°51′	88°42′	88°35′	88°27′	88°15′	88°0′	87°5′	87°5′	86°9′	84°15′	78°41′	44°15′

Note: Apparent dip (D) ＝arctan [tanA×cos (B−C)], Where A is the true dip, B is dip direction, C is strike direction, namely traverse direction.

References

[1] 石振明，孔宪立. 工程地质学[M]. 北京：中国建筑工业出版社，2013.

[2] 张忠苗. 工程地质学[M]. 北京：中国建筑工业出版社，2007.

[3] 蔡美峰. 岩石力学与工程[M]. 北京：科学出版社，2002.

[4] 王思敬等. 地下工程岩体稳定性分析[M]. 北京：科学出版社，1984.

[5] 白明洲，王勐，刘莹. 工程地质基础实习试验教程[M]. 北京：清华大学出版社，北京交通大学出版社，2010.

[6] 唐益群，石振明，黄炳炎，等. 工程地质学实习教程[M]. 上海：同济大学出版社，2012.

[7] 工程地质手册编委会. 工程地质手册[M]. 北京：中国建筑工业出版社，2007.

[8] 岩土工程师实务手册编写组. 岩土工程师实务手册[M]. 北京：机械工业出版社，2006.

[9] 中华人民共和国国家规范．岩土工程勘察规范 GB 50021—2001(2009 年版)[S]. 北京：中国建筑工业出版社，2009.

[10] 中华人民共和国行业规范．城市规划工程地质勘察规范 CJJ 57—2012[S]. 北京：中国建筑工业出版社，2013.

[11] 中华人民共和国行业规范．公路工程地质勘察规范 JTG C20—2011[S]．北京：人民交通出版社，2011.

[12] 中华人民共和国国家规范．建筑地基基础设计规范 GB 50007—2011[S]. 北京：中国建筑工业出版社，2012.

[13] 中华人民共和国国家规范．建筑抗震设计规范 GB 50011—2010(2016 年版)[S]. 北京：中国建筑工业出版社，2016.

[14] Uniform Building Code 1997[S]. Washington，D. C.：International Conference of Building Official，April 1997.

[15] International Code Council，Inc. 2003 international building code[S]. 2003.

[16] ASCE Standard. SEI ASCE 7-02 minimum design loads for buildings and other structures [S]. 2003.

[17] European Standard. Eurocode 8：design of structures for earthquake resistance. part 1：general rules，seismic actions and rules for buildings[S]. PrEN 1998-1：2003.

[18] 中华人民共和国国家规范．中国地震动参数区划图 GB 18306 —2015[S]. 北京：中国标准出版社，2016.

[19] 中华人民共和国国家规范．湿陷性黄土地区建筑标准 GB 50025—2018[S]. 北京：中国计划出版社，2019.

[20] 中华人民共和国行业规范．软土地区岩土工程勘察规程 JGJ 83—2011[S]. 北京：中国建筑工业出版社，2011.

[21] 景彦君，张以晨，周志广. 国内外对活断层的研究综述[J]. 吉林地质，2009，28(2)：1-3.

[22] 玉琴，张永兴. 欧洲抗震设计规范 Eurocode 8 简介及其与我国岩土抗震设计比较[J]. 地震工程与工程振动，2010，30(5)：134-141.

[23] 高迪，程志军，李小阳. 美国建筑技术法规简介[J]. 工程建设标准化，44-48.
[24] 褚波，宋婕. 论国内外工程建设标准体系[J]. 工程建设标准化，54-56.
[25] 高大钊.《岩土工程勘察规范》GB 50021—2001 的修订[J]. 建筑结构，2002，32(12)：62-65.
[26] 罗开海，王亚勇. 中美欧抗震设计规范地震动参数换算关系的研究[J]. 建筑结构，2006，36(8)：103-107.